W9-ACI-967

Science and Public Policy

Science and Public Policy

The Virtuous Corruption of Virtual Environmental Science

Aynsley Kellow

Professor and Head, School of Government, University of Tasmania, Australia

Edward Elgar

Cheltenham, UK • Northampton, MA, USA

Published by
Edward Elgar Publishing Limited
Glensanda House
Montpellier Parade
Cheltenham
Glos GL50 1UA
UK

Edward Elgar Publishing, Inc.
William Pratt House
9 Dewey Court
Northampton
Massachusetts 01060
USA

A catalogue record for this book
is available from the British Library

Library of Congress Cataloging in Publication Data

Kellow, Aynsley J. (Aynsley John), 1951–
Science and public policy : the virtuous corruption of virtual environmental science / Aynsley Kellow.
p. cm.
Includes bibliographical references and index.
1. Environmental policy. 2. Precautionary principle. 3. Science and state. 4. Corruption. I. Title.
GE170.K455 2007
363.7′0561–dc22

2007016149

ISBN 978 1 84720 470 7

Printed and bound in Great Britain by MPG Books Ltd, Bodmin, Cornwall

Contents

Abbreviations

ACF	Australian Conservation Foundation
AMA	Australian Medical Association
CFC	chlorofluorocarbon
CITES	Convention on International Trade in Endangered Species
DCSD	Danish Committees on Scientific Dishonesty
DD	data deficient
EBM	Energy Balance Model
EDC	endocrine disrupting chemical
EDF	Environmental Defense Fund
EWG	Environmental Working Group
FCCC	Framework Convention on Climate Change
FFI	Flora and Fauna International
FOE	Friends of the Earth
GARP	Global Atmospheric Research Program
GCM	General Circulation Model
GEF	Global Environmental Facility
GHG	greenhouse gas
GISS	Goddard Institute of Space Studies
GLP	good laboratory practice
GNI	gross national income
GSFC	Goddard Space Flight Center
GST	ground surface temperature
HIV	human immunosuppressive virus
ICSU	International Council of Scientific Unions
IGBP	International Geosphere Biosphere Program
IPCC	Intergovernmental Panel on Climate Change
IUCN	International Union for Conservation of Nature and Natural Resources
LIA	Little Ice Age
MAD	mutual acceptance of data
MAHB	Millennium Assessment of Human Behavior
MER	market exchange rates
MWP	Medieval Warm Period
NAS	National Academy of Science (USA)
NRDC	Natural Resources Defense Council

PC	principal component
PEER	Public Employees for Environmental Responsibility
PPP	purchasing power parity
SAR	Second Assessment Report
SAT	surface air temperature
SCOPE	Scientific Committee on Problems of the Environment
SRES	*Special Report on Emissions Scenarios*
TAR	Third Assessment Report
UCS	Union of Concerned Scientists
WCU	World Conservation Union
WMO	World Meteorology Organization
WRI	World Resources Institute
WWF	Worldwide Fund for Nature

Preface

The themes of this book reflect a career which started as an undergraduate in science and inevitably moved to the social sciences – anthropology and political science – and which also saw scholarship supplant and then slowly but surely replace environmental activism, an activism which saw me contest an election for the neo-Malthusian New Zealand Values Party, the first national 'green' party in the world. The inevitability of conservatism increasing with age aside, I would like to think that my basic values and desires for a decent environment have not changed, but my concern then, as now, was based upon the best possible scientific knowledge.

I no longer think that is the case. A career that has alternated between posts in environmental studies departments and political science departments has led me to the conclusion that far too much – though not all – environmental science is shot through with values, and that far too many political science analyses rather naively take on trust the fact that this is not the case. This book was motivated by a belief that, collectively, we deserve better than that.

In many ways, it is an homage to Jim Flynn, my professor at the University of Otago and supervisor (with Tony Wood) of my first piece of academic research. It is not just the quality of his teaching in political ethics, which included a detailed reading of Plato's *Parmenides* (perhaps the ultimate example of subjecting one's own work to sceptical scrutiny), but the research project that by then was beginning to consume him (and on which he generously employed me – albeit briefly – between completion of my dissertation and my first academic post).

Jim cared intensely about racism. During *his* first academic job at the University of Eastern Kentucky during the civil rights era he was imprisoned for those beliefs. But he also believed intensely in the need to produce reason and evidence to argue for political causes, and that normative beliefs alone were not enough. Jim was writing a book on the justification of humane ideals, and wanted in a few pages to provide a brief critique of the views on race and intelligence advanced by Arthur Jensen. Turning to the literature, he found there was no decent critique; instead there were numerous *ad hominem* attacks on Jensen and dismissals of his research because it had racist implications.

Jim being Jim, he wrote his own critique, published as *Race, IQ and Jensen*, at the beginning of a 20-year diversion from his book on justifying humane ideals. In the process, he not only provided the missing reasoned critique on racial differences in intelligence testing, but discovered the phenomenon, missed by the educational psychology discipline, that is now known as the 'Flynn Effect': that unstandardized IQ test results showed a steady improvement in the course of the twentieth century. (Because the psychologists were used to dealing with standardized data, they had failed to notice what an enthusiastic 'amateur' did; there are resonances in this book with the Hockey Stick case, later). For me, Jim is the paragon of what scholarship should be: important and relevant to the things we care about, but to the least extent we can manage, uncontaminated by them.

Ironically, this book was commenced while I was back at Otago on sabbatical on an Edward Wilson Fellowship, resident at St Margaret's College, and it owes much to Otago both then and in my days as a student, both to Jim and his colleagues in the Department of Political Studies, including Marian Simms, who now holds the chair and with whom I was supposed to be working on interest groups, but to those in the Philosophy Department under Alan Musgrave. For many years the Otago philosophers provided not just interesting discussions of the philosophy of science, but also a constant stream of interesting visitors (including Karl Popper). Particular thanks go to John Norris, Warden of St Margaret's, and his residents for taking an expat New Zealander into their bosom during the fellowship, and to the University of Otago for its support. Similarly, I am indebted to the University of Tasmania for supporting the sabbatical, and especially to my colleagues in the School of Government for enduring several seminars where I first explored the subject matter of the first three chapters. Others, from outside the School provided useful comments, including scientists in the Antarctic Climate and Ecosystems Cooperative Research Centre, and the School of Zoology, especially Randy Rose. Rob Hall and Marcus Haward (both ACE CRC and School of Government) read drafts and provided assistance in many ways, and Ted Lefroy read and commented on the entire manuscript and suggested many improvements. Ian Castles, subject of part of the book, also provided many useful details and read some early fragments, providing invaluable comments. Other useful comments, too numerous to catalogue individually, were received on fragments of the argument at two National Academies Forums and an Australian Academy of Science annual symposium in Canberra. Finally, John Adams has for many years proved a wonderful source of knowledge and inspiration on all matters related to risk and the policy uses of science, and Ted Lowi gets yet another citation in one of my books. (Somehow, wherever my scholarship leads me, Ted has been there and left at least some slight footprints!) As always, any errors are mine.

And, as always, Julie and Maddy have tolerated the long hours with my nose buried in what has proven to be the false promise of the paperless office. Julie has been burdened with enthusiastic descriptions of various points to an extent that would try the patience of a saint, while Maddy is still too young to be harangued. But it is to her, in the hope that she will inherit a world in which there is both beauty in truth, and truth in beauty, that this book is dedicated.

1. The political ecology of *Pseudonovibos spiralis* and the virtuous corruption of virtual science

> They hunted till darkness came on, but they found
> Not a button, or feather, or mark,
> By which they could tell that they stood on the ground
> Where the Baker had met with the Snark.
>
> In the midst of the word he was trying to say,
> In the midst of his laughter and glee,
> He had softly and suddenly vanished away –
> For the Snark was a Boojum, you see.
>
> Hunting of the Snark, **Lewis Carroll**

In late 2003 the International Union for Conservation of Nature and Natural Resources (IUCN) released a new edition of its Red List of endangered species. On the list, again, was *Pseudonovibos spiralis* (Peter and Feiler, 1994), commonly known as the *khting vor* ('wild cow with horns like lianas') in Cambodia or the *linh duong* ('mountain goat') in Vietnam. *Pseudonovibos spiralis* had been first listed in 2000 (IUCN, 2003).

Pseudonovibos spiralis was categorized as 'EN C2a' under the categories and criteria adopted by the IUCN in 1994, meaning it was 'endangered', with a population estimated to number less than 2500 mature individuals, and suffering a continuing decline 'observed, projected, or inferred' in numbers of mature individuals and population structure in the form of severely fragmented subpopulations, each of no more than 250 mature individuals.

That the Red List continued to include *Pseudonovibos spiralis* was perhaps surprising, but even more surprising was the fact that the IUCN, if it had to list it, did not include the designation 'DD', or 'data deficient'. This category is used when 'there is inadequate information to make a direct, or indirect, assessment of its risk of extinction based on its distribution and/or population status'. Clearly, the IUCN felt there *was* sufficient data upon which to make an assessment – direct or indirect – of the risk of extinction of *Pseudonovibos spiralis.* This would have surprised many scholars, because there is every indication that *Pseudonovibos spiralis* was more than extinct – that it had not just ceased to exist, but that it *never had existed.*

While several of the world's museums contain purported specimens of *Pseudonovibos spiralis*, many scientists dismiss them as fakes. All of the 21 specimens in existence have been collected from shops and markets in Indo-China (Whitfield, 2002). Other than its discoverers, no zoologist has actually seen a *Pseudonovibos spiralis*, but there is a mythical beast in Cambodian folklore, a snake-eating cow with twisted horns (the *khting vor*) which believers assert is *Pseudonovibos*, and a Chinese encyclopaedia from 1607 describes a beast which sleeps by hanging from a tree by its horns, which doubled as both an unusual means of sleeping and a cure for snakebite. This, too, is supposed to have been inspired by *Pseudonovibos.*

The case of *Pseudonovibos* provides some revealing insights into the relationship between contemporary science and environmental politics, not just in the area of conservation biology, but more broadly, where much of the science in areas from biodiversity to climate change has become virtual rather than based on observation and experimentation – based, in other words, on inference, extrapolation and mathematical modelling of what *might* or *could* be the case. Much environmental science relies upon the availability of computing power not available to past generations of researchers, and the words 'might' and 'could' are now commonplace, and frequently used to connect virtual research to some perceived problem, and thus to a political agenda.

This chapter explores this problem with reference to *Pseudonovibos.* It does so by first examining the facts of the *Pseudonovibos* case and then discussing some related issues in the conduct of science, particularly its politicization which, it is argued, frequently results in its 'virtuous corruption' – what is known in law enforcement circles as 'noble cause corruption' (where, 'knowing' a suspect is guilty, police officers manufacture evidence to ensure a conviction). The expression 'noble cause corruption' refers to situations where it is supposed that 'if the goal was thought to be worthy, the means did not matter very much' (Kleinig, 2002, p. 301). It involves something more serious than a 'little white lie', because the untruths are often substantial, but the distinguishing characteristic is that it involves the invocation of some virtuous cause. In the case of policing, it typically involves perjury in the form of 'fitting up' a suspect one 'knows' is guilty by fabricating evidence of guilt to ensure that a suspect who 'deserves' to be found guilty does not slip through the net of the justice system.

The word 'corruption' does not just mean 'the perversion of integrity by bribery or favour' (*Shorter Oxford English Dictionary*), which is perhaps now its most common contemporary usage. It also refers to 'the destruction or spoiling of anything especially by disintegration or decomposition'. Something which is corrupt is said to be 'changed from the naturally sound condition', or (especially in relation to language) 'destroyed in purity, debased'. It is in this sense – captured equally well with the sense of corruption of a computer disk – that corruption is meant here, but the term is used in full knowledge of the irony, in writing of virtuous corruption, that it also has usage related to the bringing about of such debasement through venal and depraved means. Since we focus here on noble cause corruption in relation to the conduct of virtual science, the parsimony and alliterative qualities of 'virtuous' are attractive, but there is also more than just alliteration in the choice of words: 'virtual' and 'virtuous' have a common derivation from the Latin word 'virtus', meaning 'force, potential'. (We might also note that 'virtus' derives from 'vir', meaning 'man', and that an overwhelming majority of the scientists I study here are men, but I leave to others the possibility of an analysis of the gendered nature of contemporary science).

This book will argue that the virtual nature of much environmental science and the application of non-science principles such as the precautionary principle, facilitate the virtuous corruption of environmental science. While drawing upon examples from conservation biology and biodiversity, it will also suggest that the problem is more widespread than this area alone would suggest, and is common in the important field of climate science. It will argue for the importance of reliable science as the basis for environmental management and policy. As a side observation, it will also suggest that a purely 'scientific' basis for public policy is a chimera: there is rarely a linear relationship between science and public policy, with scientific understanding leading to only one policy option. First, however, we shall consider in greater depth the case of the *khting vor*, because it provides insights into the nature of the virtuous corruption of science.

THE CURIOUS CASE OF THE *KHTING VOR*

Pseudonovibos was first described to science in 1994 by two researchers from Dresden, Germany, who described a creature near the Vietnam–Cambodian border (Peter and Feiler, 1994). They apparently undertook no field research; rather, they claimed to have discovered the *khting vor* after collecting a set of horns and unearthing others in various markets in the area (Hespe, 2002). Hundreds of Khmer hunters and villagers claimed sightings of *Pseudonovibos* in studies conducted after the German report, and the species programme manager for the Worldwide Fund for Nature (WWF), Lic Vuthy, was convinced that the German report confirmed the stories of the power of the *khting vor* which had been passed on by his family for generations.

The political situation in Cambodia aided the credibility of reports that a large, previously unrecorded ungulate might exist in the forests near the Vietnamese border, as until the Khmer Rouge were ousted from the area, only in 1997, large tracts of forest had gone unexplored, at least by Western scientists.

Two sets of horns in the collection of the University of Kansas Natural History Museum, which had been found in Vietnam in 1929 and labelled as a kouprey (*Bos sauveli*, a large wild member of the cattle family, itself unknown to science until 1937), were subsequently claimed to constitute hard evidence of *Pseudonovibos spiralis* after microscopic examination and DNA testing suggested it represented a new species (Timm and Brandt, 2001; Gee, 2001). These horns had been collected by a father-and-son team of big-game hunters, Richard L. Sutton and Richard L. Sutton Jr, who had collected the horns from oxen killed for meat and tiger bait while bagging elephants and tigers 125 miles north-east of Ho Chi Minh City in January 1929. The horns were then presented to the University of Kansas in 1930. While Timm and Brandt were convinced the horns were specimens of *Pseudonovibos spiralis*, their reported collection point was much farther south than the range reported by Peter and Feiler.

The University of Kansas specimens were claimed to be the most complete, best documented and oldest specimens of *Pseudonovibos.* They were believed to comprise a male and a female. But the most complete set consisted only of horns and a few skull parts – the posterior half of the frontal bones, the parietals, the horn cores and horns, and the anteriormost supraoccipitals. Timm and Brandt drew upon traditional references to *khting vor* from the 1880s and 1950s, with references to its putative magical powers over snakes, to strengthen their claim that the University of Kansas specimens were *Pseudonovibos.* Henry Gee reported in *Nature* that Timm and Brandt found great significance in the observations of French civil engineer Edgar Boulangier, who had visited Cambodia for six months in 1881. Boulangier reported a species of wild ox called a *khting-pos* which supposedly fed on snakes, giving its horns talismanic powers against snakebite. Because present-day Khmer hunters have similar beliefs about *Pseudonovibos*, Timm and Brandt thought Boulangier must have been describing *Pseudonovibos* (Gee, 2001). Worryingly absent from all of this physical evidence and 'traditional knowledge' were any sightings or photographs by scientists of a live beast, any spoor or any bones other than skulls and horns which might be traded in markets. If the University of Kansas specimens were *Pseudonovibos*, it was, of course, possible that it had subsequently become extinct, which might explain the absence of other corroborating evidence.

Pseudonovibos had by this stage begun to infiltrate the literature of intergovernmental organizations and nongovernmental organizations. In August 1999, Kristin Nowell of Cat Action Treasury reported to David Phemister of the Save the Tiger Fund of the National Fish and Wildlife Foundation, Washington, DC on progress on CAT's Cambodian Tiger conservation programme, and requested payment of the final instalment of their grant. Citing an example of the importance of their work on tiger prey, Nowell referred to a chapter on the *khting vor* prepared for the IUCN Asian Antelope Action Plan and a report on the status of wild cattle in Cambodia, in which the *khting vor* even received billing in the title (Nowell, 1999).[1] A story in the *Phnom Penh Post* in 1998 on one of the authors of the status report, Hunter Weiler, reported on his tiger conservation efforts in the Cardomom Mountains, where he had seen wandering herds of elephants and wild cattle such as bantengs and gaurs. But, it reported, there were also many other rare species – 'including the reclusive gazelle-like *khting vor*' (Grainger, 1998). It was as if everyone *wished* there was a *khting vor*, but there was no hard evidence. Indeed, surveys published in 1996 by Flora and Fauna International (FFI), WWF and IUCN-World Conservation Union in Eastern Cambodia had failed to find any evidence of *Pseudonovibos*, though its presence was 'not ruled out conclusively'. Instead, a survey was recommended to establish its presence or absence, though the report read as if it was only its presence in Mondolkiri and Rattanakir provinces that was in doubt (Desai and Vuthy, 1996).

Pseudonovibos was therefore taking on an existence of its own, on the basis of folklore and the Peter and Feiler paper, in the context of the effort to save the tiger, and ultimately campaigns to preserve tropical rainforest and to ban landmines. This last linkage was due to the impact of landmines in the area formerly occupied by Khmer Rouge forces not just on human, but also on animal populations. (Given the extensive landmine hazard, one might have expected the finding of more specimens, albeit deceased ones.) *Pseudonovibos* had begun to acquire iconic status. Then in 2000, the UK NGO Global Witness claimed that a survey of the Cardamom Mountains by Fauna and Flora International had found 'evidence of threatened Asian mammals such as tiger, elephant, gaur, banteng and the khting vor' (Global Witness, 2000, p. 11). Global Witness cited newspaper reports in the *Phnom Penh Post* and *New York Times* as its authorities for this statement. FFI was somewhat more restrained in its own reports on the success of this survey, referring only to 'more reports' of *khting vor* in one report (WIT, 2000), although its website documenting its Asia-Pacific programme referred to 'remarkable densities of wild cattle – possibly including the mysterious *khting vor*' (FFI, 2003).

WWF included *Pseudonovibos* (the 'enigmatic' *khting vor*, 'known to science only by a few horns') among the 'impressive large vertebrates' in its *Wildworld Profile of the Southeastern Indochina Dry Evergreen Forests* (WWF, 2003), and it made its way into Global Environmental Facility (GEF) projects. For example, one study referred to Viet Nam's global biodiversity significance, which had been highlighted by the discovery in the past four decades of four large mammal species, including the *khting vor* (interestingly, using the Khmer common name, rather than the Vietnamese *linh duong*) (Casellini *et al.*, 2001, p. 1). *Pseudonovibos* was also mentioned in a concept paper for a UN Development Program project under GEF auspices for a 'Conservation Area through Landscape Management' for the Northern Plains of Cambodia, for example.

Pseudonovibos had quickly become woven into the international politics of tiger conservation and rainforest conservation. Like several other endangered species, *Pseudonovibos* provided a justification for preserving habitat, and for committing funds for the same from the GEF and other sources. Nobody really had a strong interest-driven reason for disputing its existence, but some scientists had begun some sceptical research.

First, a group of Austrian and German scientists took DNA from the Dresden specimens and published a paper in 1999 in which they stated it appeared to be closely related to sheep and goats (Whitfield, 2002). Then Russian scientists published results of DNA analysis of different specimens in 2002 suggesting it was a type of buffalo. In the absence of hard data or reliable observations, these contradictory findings from DNA analysis sounded alarms bells for some.

A French team then entered the fray with damning evidence that *Pseudonovibos* was a fake. Arnoult Seveau of the Zoological Society of Paris and two colleagues, palaeontologist Herbert Thomas of the College de France and biochemist Alexandre Hassanin of Paris-IV University, examined four sets of horns bought at marketplaces in Cambodia and Vietnam and concluded that they were clever forgeries (Hassanin *et al.*, 2001). They had searched forests and meat markets for six months without finding any proof that a real animal had ever existed, encountering only the usual myths and legends. DNA analysis showed the skull bones had come from a common cow which had been pieced together with horns that had been heated, moulded and carved. Their research method included examining silicon rubber moulds of the inside of the horns and microscopic examination of the outside. The apparent goat DNA sequences in the earlier analysis were attributed to laboratory contamination with chamois DNA. The French team argued that *Pseudonovibos* was a cash cow for local traders who sold the horns as cures against snake bites.

The dispute was addressed by a debate in the pages of the *Journal of Zoology* between the French team, the University of Kansas team and two others, each on different sides (Brandt *et al.*, 2001). Richard Melville, a scholar with 40 years of experience in Cambodia, was convinced the horns were fakes and he labelled the believers' claims 'tortured conjecture'. He considered the manufactured horns were accredited with mystical powers because of the centrality of snakes to the mixture of Hindu and Buddhist elements in the Cambodian cultural tradition. Researchers had been misled, according to Melville, by their cultural ignorance.

Within the IUCN, inclusion of *Pseudonovibos* on the Red List did not excite opposition, and the lack of evidence was no impediment to listing. There were numerous motivations for listing it, and an absence of any strong reason for any actor to question its listing. *Pseudonovibos* had been assessed by the IUCN/SSC Antelope Specialist Group for inclusion on the Red List in 1996, a time when only two papers by Peter and Feiler had appeared in the literature. A single paper by the Dresden authors was cited as the Species Authority. None of the papers containing the conflicting DNA analyses, nor the analyses of the French team had been incorporated in the original assessment, and they did not dissuade the IUCN from continuing to list *Pseudonovibos* in the 2003 List. Rather, the 2003 list contained the following justification:

> The existence and systematic position of *Pseudonovibos spiralis* is currently being debated. There are undoubtedly manufactured trophies ('fakes') in circulation, but the precautionary principle requires us to assume that the species did exist and may still exist.

Pseudonovibos spiralis was still on the IUCN Red List as late as December 2006, despite its scientific epitaph having been written. The owner of the largest private collection of Asian ungulate trophies, Prasert Sriyinyong, reported that Cambodians have long been manufacturing these folklore-related items, and even described the method: buffalo or domestic cattle horns have their sheath removed, are soaked in vinegar, heated in sugar palm and bamboo leaves, given a twist at the tip, and impressed or scored to create annulations (Galbreath and Melville, 2003, p. 169). As two sceptics put it:

> In sum, it is time to bring to an end the saga of the spiral horned ox *Pseudonovibos.* Anyone wishing to challenge the null hypothesis that no such creature has ever existed in South-east Asia should provide a specimen that can be subjected to definitive tests pertaining to possible alteration, and which passes such tests. To date, no such specimen has emerged. (Galbreath and Melville, 2003, p. 169)

Moreover, it now seems the kouprey has dubious status as a species, with genetic testing suggesting it is a wild hybrid (Galbreath *et al.*, 2006).

VIRTUOUS CORRUPTION AND VIRTUAL SCIENCE

The case of the *khting vor* provides an almost comical example of the operation, jointly, of consensus 'science' and the precautionary principle in international politics. Much could be written of the process whereby the IUCN consensus (or other international consensus documents on science) was produced, but suffice it to say that nobody really had a strong reason to oppose its inclusion, and plenty had some reason to list it. For any sceptics, the invocation of the precautionary principle has been enough to repel dissent. After all, it *might* have existed, and the absence of any confirmed sightings, photographs or hard physical evidence surely suggest that (if it did) it is indeed now extremely rare and endangered. The international deliberative processes followed by the IUCN and other arbiters of scientific consensus merit further discussion, but we shall confine ourselves here to looking at 'virtuous corruption' and what it means for the conduct of science, particularly the environmental science which underpins much contemporary environmental policy.

The case of the *khting vor* is strictly not just one of virtuous corruption of science, of evidence being fabricated. Rather, it is also a case of the 'jury' of the IUCN reaching a decision on the basis of very flimsy evidence indeed. To some extent, the fabrication of evidence of the existence of the *khting vor* was probably conducted by those other than the scientific community. We will examine some other cases in a later chapter where evidence appears to have been fabricated, but the *khting vor* case illustrates the seductive power of a noble cause in deliberative processes as well as in the conduct of science.

We are more used to thinking of corruption in science, of scientific fraud and the like, as occurring as a result of more base motives – as the result of the influence of strong commercial interests. While not diminishing the importance of this kind of corruption, we can add to those possibilities the reputational motives that lead some scientists to fabricate evidence or commit other kinds of fraud. Often, we might reduce such motives to economic ones, as scientists seek research grants or career advancement, but the importance of reputation alone should not be ignored. The pressures can be strong, especially when institutions depend upon research success for financial security. Money is most apparent as a corrupting influence with medical research, where pharmaceutical companies stand to make considerable profits on the basis of product approvals which depend upon evidence of both efficacy and safety.

But even in the area of drug research, noble causes can also seduce researchers into fabricating evidence in later research when faced by accusations of fraudulent behaviour. For example, Australian Dr William McBride, the medical scientist who first detected the problems with the drug thalidomide, later stood accused of fabricating evidence when faced with the need to generate funds to support the institute he founded to research birth defects. And this example underscores one of the problems with virtuous corruption: false science has costly consequences, because it might lead to the benefits of otherwise safe and efficacious drugs being lost to humanity, or in dangerous ones being approved. Medical research has developed numerous processes to guard against fraud and poor research. Because of the importance of intellectual property rights, much research upon which drug registration decisions is based is conducted in-house in the research laboratories of the drug companies or by independent laboratories on a contract basis. To deal with the obvious problems of venal corruption, research must be conducted according to strict protocols, with the practice of research laboratories subject to auditing by regulatory authorities. In addition, the research is usually conducted using 'double-blind' approaches: those administering doses do not know actual doses from placebos, and diagnosis and analysis is also conducted by distinct teams. All this minimizes the chance of even expectations, let alone interests, from intruding into the research process.

Then, publication of results in scientific journals occurs only after anonymous peer review, where the identities of the authors and reviewers are not known to each other, so that the focus is solely on the science, rather than on reputation or other factors. Commonly, with medical journals, a declaration is made of any competing interests, so that readers can be made aware of any possible biases. Most do not, however, require declarations of support for noble causes such as better human health or environmental protection, and there is less recognition of the problems of virtuous corruption than in the area of law enforcement.

The argument here is that this corruption in the name of a noble cause is facilitated by the virtual nature of much environmental science. By this is meant a reliance upon mathematical models, particularly large complex computer models. The most notable recent example of such models was arguably the neo-Malthusian *Limits to Growth* study published by the Club of Rome in 1972, which suggested the world would run out of resources. The debate following *Limits to Growth* saw frequent use of the computing acronym GIGO – 'garbage in, garbage out' – but what is less widely appreciated is that ecological science is replete with the use of such techniques.

Moreover, some of the major issues of our time, such as global warming, depend similarly upon 'virtual research'. Rather than being based upon observational data, much of our understanding is based upon the results of computer models, and much of the observational data fed into them is in turn either virtual in nature (including proxies rather than direct evidence) or the result of a worrying degree of manipulation which leaves open the possibility of bias.

Both climate science and the science of biodiversity depend heavily upon virtual science, and when both are combined, the problems are compounded. For example, in January 2004 a paper published in the leading journal *Nature* combined the virtual science of climate change with the virtual science of biodiversity as it warned of the loss of thousands of species with a relatively small warming over the next century. But just how virtual was this science is apparent when we consider that the estimates of species loss depended upon a mathematical model linking species and area; modelled changes in the areal distributions of areas of habitat depended in turn upon the results of climate models tuned to reflect climate changes as a result of increasing greenhouse gases (GHGs) such as carbon dioxide; these in turn were driven by scenarios of what GHG emissions might look like over the next century, driven in turn by economic models. Sceptics were quick to point out the warming over the previous century had not left any such trail of species devastation, suggesting a disjuncture between the real world and the virtual world, but this was virtual biological data, generated by virtual climate data, driven by virtual emissions data, driven by virtual economic data.

This gives some idea of just how far contemporary environmental science published in the leading journals has come from notions of science driven by observational evidence. Often, such 'science' is not conducted with the same kind of safeguards found with medical research: the same scientists collect, manipulate, analyse and interpret the 'data', and the opportunities for either deliberate fraud or inadvertent manipulation as the result of subjective factors are thus great. We accept that there are problems such as biodiversity loss and climate change, but the nature of the science underpinning them is worrying because it is rarely determinate and its virtual nature leaves ample scope for the scientists' subconscious to affect its conduct.

SCIENCE ABDUCTED BY ALIENS?

Novelist Michael Crichton, a medical practitioner who has expressed concern at the differences in standards between medical research and other areas of research, has suggested the onset of largely virtual science commenced with the foray into the 'science' of the search for extraterrestrial intelligence (Crichton, 2003). In 1960, a young astrophysicist named Frank Drake ran a two-week project to search for extraterrestrial signals. There was great excitement when a signal was received, but it turned out to be a false positive. Undeterred, Drake organized the first 'Search for Extraterrestrial Intelligence' (SETI) conference, and developed what became known as the Drake equation:

$$N = N^* f_p \, n_e f_l f_i f_c f_L$$

where:

N is the number of stars in the Milky Way galaxy;
f_p is the fraction with planets;
n_e is the number of planets per star capable of supporting life;
f_l is the fraction of planets where life evolves;
f_i is the fraction where intelligent life evolves;
f_c is the fraction that communicates; and
f_L is the fraction of the planet's life during which the communicating civilizations live.

Most of us are probably familiar with the conclusion derived from the Drake equation which penetrated popular consciousness – that there are so many stars in the Milky Way that, 'statistically', there *must* be intelligent life somewhere. As Crichton pointed out, this was 'pure speculation in quasi-scientific trappings', because none of the values can be known and must be just guessed at, and the result is not falsifiable. Yet despite its problems, the Drake equation was largely tolerated by science as a curiosity. There were many critics, but there were few among the ranks of astrophysicists and astronomers, perhaps because it captured the public imagination and resulted in higher political support for their research budgets.

But, as Crichton pointed out, this speculation disguised as science by the cloak of sophisticated mathematics set the scene for more serious speculation in the 1970s and 1980s, this time in the name of a laudable objective – world peace in a nuclear age. This was the creation of the alarm over what was known as 'nuclear winter' in the 1980s, which Crichton sees as a forerunner to the substantially virtual science of climate change in the 1990s.

The US National Academy of Sciences had reported in 1975 on the effect of dust from nuclear blasts, but found it to be relatively minor. In 1979, the Office of Technology Assessment issued another report concluding that, while nuclear war could *perhaps* have irreversible adverse consequences for the environment, the scientific processes involved were poorly understood and it was not possible to estimate their probable magnitude. Notwithstanding these reports, in 1982, the Swedish Academy of Sciences commissioned a report (Crutzen and Birks, 1982) entitled 'The Atmosphere after a Nuclear War: Twilight at Noon', in which the authors speculated that there would be so much smoke that a large cloud would cover the northern hemisphere for several years and reduce incoming sunlight to levels below those required for photosynthesis.

Then in 1983, five scientists (including popular cosmologist Carl Sagan) published a paper in the leading journal *Science* that attempted to quantify the atmospheric effects of a nuclear war, with credibility added by the use of a computer model of global climate – though, significantly, they constructed their own model specially for this purpose rather than use the very early climate model that was by then available (Turco *et al.*, 1983). At the heart of this undertaking was another equation, never specifically expressed, but one that could be paraphrased as follows:

$$D_s = W_n\, W_s\, W_h\, T_f\, T_b\, P_t\, P_r\, P_e \ldots \text{etc.}$$

(the amount of tropospheric dust=number of warheads×size of warheads×warhead detonation height×flammability of targets×Target burn duration×Particles entering the Troposphere×Particle reflectivity×Particle endurance . . . and so on).

Crichton saw this line of 'science' as leading directly to the science of climate change, and there is some foundation to his argument, with many of the proponents of anthropogenic climate change first emerging in the literature surrounding nuclear winter.

But Crichton overlooked another paper in the same 1983 issue of *Science* by Sagan and Paul and Anne Ehrlich and 17 others, including Bob May – later to rise to the position of Chief Scientist in the United Kingdom and President of the Royal Society (Ehrlich *et al.*, 1983). This paper suggested the consequences of a large-scale nuclear war would be subfreezing temperatures, low light levels, and high doses of ionizing and ultraviolet radiation extending for many months, which could destroy the biological support systems of civilization, in the Northern Hemisphere at least. The productivity of natural and agricultural ecosystems might be severely restricted for a year or more, and survivors would face starvation as well as dark, freezing conditions while exposed to near-lethal doses of radiation. The Southern Hemisphere might also be affected directly, but there would be indirect effects because of the interdependence of the world economy. The extinction of a large proportion of the Earth's animals, plants and microorganisms seemed possible, and it was thought that the population size of *Homo sapiens* could be reduced to prehistoric levels or below, with the possibility of the extinction of the human species itself unable to be excluded.

Nuclear winter appears, therefore, to be a forerunner not just of the science of climate change, but of the science of the impact of climate change on biodiversity which was to be a feature of the concern over climate change from the 1990s. It also represented a continuation of the neo-Malthusian concerns that Ehrlich, May and others had raised in the late 1960s and early 1970s, but had been legitimated earlier still by modelling in the science of biodiversity. We shall return to the case of nuclear winter in Chapter 5, and make some further remarks about the importance of this early example of politicized science, and how closely it anticipated certain features of climate science (such as the use of the worst possible assumptions to produce politically useful conclusions). But we shall conclude this chapter by noting that the phenomenon of virtuous corruption is by no means confined to the cases we shall examine here of biodiversity and climate change.

VIRTUOUS CORRUPTION: HOW WIDESPREAD?

There is a growing list of candidate cases of virtuous corruption of environmental science. A paper suggesting contamination of native species of maize by genetically modified maize was formally withdrawn by the editors of *Science* after flaws in its methodology were exposed. A similar case involving supposed synergistic reactions amplifying greatly the effects of endocrine disrupting chemicals (EDCs) saw formal withdrawal after failed attempts at replication. In this case, funding for the study (at Tulane University) had come from the W. Alton Jones Foundation, the Director of which was a co-author of the book which popularized the EDC issue, *Our Stolen Future*, and which provided substantial grants to environment groups to raise awareness of the problem of EDCs. By the time the peer-reviewed paper had been withdrawn, however, the US Congress had mandated the Environmental Protection Agency to regulate EDCs. (We shall also look at these cases in more detail in Chapter 5.)

There are similar examples in medical science, such as the alarm over the MMR vaccine in the UK on the basis of a flawed study involving fewer cases than analytical categories, which facts alone should have alerted referees to the limitations of the study. While medical journals are usually more exacting in their standards than many journals in other disciplines, and the editors of *The Lancet* later stated the study should not have been published because of a serious conflict of interest involving the chief investigator, the system was undone by the desire to protect vulnerable children. The story of silicone breast implants is similar, with multi-billion dollar lawsuits driving Dow Corning into bankruptcy before definitive studies demonstrated no evidence that implants caused most of the maladies claimed. Similar false alarms have been raised over cancer and powerlines and mobile telephones.

But part of the problem with biodiversity science which facilitates virtuous corruption is the virtual nature of much of the science, and areas where there is a similar virtual component are where virtuous corruption is most likely to occur. Climate science is perhaps the most significant area of environmental science where the virtual nature of the science allows many possibilities. Leaving aside the validity of the General Circulation Models (GCMs) which run on supercomputers in an attempt to simulate the highly complex global climate systems, the crucial inputs to these models – such as future levels of CO_2 – are generated by scenarios which are no more than 'what if?' exercises. Scenarios should never be regarded as forecasts which, regardless, are notoriously difficult to construct with any accuracy over the 20-odd years which form the basis for the planning environments of electric utilities, yet alone the 100-year period in most climate models (see Kellow, 1996). The release of a Pentagon scenario exercise in February 2004 demonstrated all too well the willingness of political actors to abuse such virtual science. The Pentagon study quite explicitly involved a 'what if?' scenario of a naturally-occurring set of circumstances 8200 years ago, but Greenpeace had no hesitation in claiming that this was somehow a forecast of future anthropogenic climate change.

There are similar examples from the science surrounding climate impacts. There are claims that global warming will cause an increase in 'tropical' diseases like malaria. This science ignores both the fact that malaria is not tropical (it was once endemic up to the Arctic Circle) and the fact that socio-economic factors are mainly responsible for its eradication in Europe and North America. Climate science takes the results from GCMs that warming will result in a wider distribution of insect disease vectors to suggest that insect-borne diseases will become more common, yet it ignores the fact that those models are driven by socio-economic scenarios which assume that those whose less fortunate socio-economic circumstances currently result in higher infection rates will become more like those whose affluence now allows them protection through air conditioning, insect screens, better medicine, and so on. This highly reductionist science assumes greater wealth will result in higher emissions, but none of it will be spent on the things that make disease transmission much less likely.

CONCLUSION AND PROSPECT

The context within which environmental science is conducted contains numerous factors which might facilitate its virtuous corruption, especially where the science is largely virtual – that is, relying primarily upon the results of mathematical models, be they simple ones or those which rely upon substantial supercomputing, rather than more concrete prediction and observation. This is especially so because data usually must be manipulated substantially in order to be usable, and this provides scope for either deliberate 'massaging' or the unintentional intrusion of subjective beliefs.

Cultural factors undoubtedly affect both the conduct and interpretation of science, but there are particular aspects of environmentalism that enhance this possibility. There is a theme in environmental thought that rejects the rationalism of the Enlightenment and suggests traditional ways of knowing are unduly rejected by modern science. Ironically, witchcraft (often celebrated as one such 'traditional way of knowing') is related to one manifestation of contemporary environmentalism, which was important in the *khting vor* case: the precautionary principle. The precautionary principle is a contested concept, but it is often interpreted by environmental groups as requiring proof of safety – a proposition (the proof of a negative) which is equally impossible as the demand that those accused of witchcraft before the Enlightenment prove their innocence. Regardless of what one thinks of the merits of the precautionary principle, and how it might be operationalized, the examples of virtuous corruption discussed here suggest that in the speculations of virtuously corrupt science (typified by the inclusion of words like 'could' and 'might') lie the dangers of 'witchcraft' tests that can never be satisfied. Interestingly, anthropologist Mary Douglas (1992) has suggested that many accusations of risk serve the same function as witchcraft in pre-modern societies: the attribution of blame to attack powerful corporations and other actors.

Such rejections of the rationalism of science, with all its disciplines, carry with them dangers. Traditional ways of knowing not only detract from the good science which must underpin any sound environmental management (including wildlife conservation), but the ways of traditional knowledge can actually work against conservation biology and facilitate species extinction. It is, after all, not the rationalism of Enlightenment science which leads to a demand for tiger penis and other products of endangered species, but traditional folk medicine. To tolerate the corruption of science – even for virtuous purposes – is thus one thing which works against good environmental outcomes. Good environmental outcomes depend crucially on the conduct of the best environmental science conducted by scientists who are on tap, not on top (as C.P. Snow once famously put it).

This book has the following structure. First, I examine virtuous corruption in the areas of conservation biology (next chapter) and climate science (Chapter 3). Then I use the reaction of many of the practitioners of virtual science to the publication of Bjorn Lomborg's *The Skeptical Environmentalist* to explore the politics surrounding the conduct of environmental science, before turning (in Chapter 5) to discuss the extent to which politics and social factors are currently impacting upon the conduct of science. It has become fashionable to refer to a 'war on science' being waged by the Republican Party in the United States, but (as we shall see) this is not a one-sided contest, and it is by no means clear that the Republican Administration of George W. Bush fired the first shot. What is clear is that science and public policy-making based upon it are much the poorer for the conflict, and we would do well to provide better quality assurance in the provision of science for policy-making. Finally, I attempt some theoretical explanations for the virtuous corruption of science under conditions where there is considerable reliance upon models and simulations, running on highly massaged data, rather than more conventional science of testing hypotheses against observational evidence.

I will conclude by suggesting that this virtuous corruption of virtual science derives from divergent ontological positions – that the perpetrators carry with them different theories of nature and human nature, and they are often unaware that their 'science' embeds a different political philosophy, so that they simultaneously accuse their opponents of politicizing science, all the time remaining ignorant of the extent to which their own science is politicized.

NOTE

1. The status report was Kimchhay *et al.* (1998).

2. The political ecology of conservation biology

> Where we have strong emotions, we're liable to fool ourselves.
>
> **Carl Sagan**

The case of the *khting vor* provides an almost comical example of the operation, jointly, of consensus 'science' and the precautionary principle in international politics. One can find other examples where the 'consensus science' of international organizations leads to curious decisions.

In its *Global Environment Outlook 3* the UN Environment Program claimed that 'dead zones' had recently been appearing off the coasts of New Zealand, southeast Australia, Japan, China and South America. 'Dead zones' in the seas and oceans were thought to be caused by an excess of nutrients – mainly nitrogen – from agricultural fertilizers, vehicle and factory emissions and wastes. The resultant low levels of oxygen in the water make it difficult for fish and other marine creatures to survive.

Two of these dead zones were off the coast of New Zealand. Surprisingly, one was off the coast of Fiordland, a remote wilderness area one would not think would be a source of nutrient run-off. Publication of the *Year Book* produced something of a mystery in New Zealand, because nobody seemed aware of any nutrient problems. Dr Janet Grieve, a biological oceanographer with the National Institute of Water and Atmospheric Research, was quoted by the *Dominion Post* newspaper as saying that she was not aware of any oxygen-starved zones off New Zealand that would fall into the 'shock-horror' category, and believed the report was somewhat misleading.

The international deliberative processes followed by the arbiters of scientific consensus that led to the listing of *Pseudonovibos* merit further discussion, but we shall confine ourselves in this chapter to examining other examples of virtuous corruption in the area of conservation biology and other ecological sciences and what they mean for the conduct of science. From these we can begin to draw some observations about the relationship between science and the context within which it is conducted.

It is not just international processes that facilitate errors in the name of a virtuous cause such as the protection of mythical creatures. For example, in 1986 the regional council in the north-western Swedish province of

Jamtland placed on a list of endangered animals a serpent-like creature believed to inhabit Lake Storsjon. Its endangered species listing was lifted in 2005, making it open season for any hunter who could find the beast. Sweden was not alone in its concern for its mythical creatures. The Scottish Office in the UK debated protection for the Loch Ness monster prompted by an approach by the Swedish government to the British Embassy in Stockholm in 1985 seeking advice regarding protection for the Lake Storsjon creature. After some light-hearted memoranda, in which it was pointed out that Nessie was not a salmon, and therefore not covered by the Salmon and Fisheries Protection (Scotland) Act 1951, it was decided Nessie should be protected under the Wildlife and Countryside Act 1981, which made it an offence to snare, shoot or blow it up (*The Herald*, 9 January 2006).

A more pervasive problem of virtuous or noble cause corruption can be seen when we turn to conservation biology and the concerns over biodiversity loss. Greenpeace, seeking to alarm the public and solicit donations, often cites a figure in its advertisements of 50 000–100 000 species becoming extinct every year. Yet others have put the documented extinction rate as low as a single species per year. Jeff McNeely, chief scientist at the IUCN-World Conservation Union (WCU), acknowledged on the eve of the Eighth Conference of the Parties to the Convention on Biological Diversity that only around 1.7 million plant and animal species had been described, but some estimates were that there might be 100 million species. The gulf between the numerous estimates of extinction numbers (unable to be described as rates without an estimate of total numbers) and actual documented extinctions was even wider, with the WCU estimating there had been only 'more than 800' plant and animal extinctions since 1500 when accurate historical and scientific records began' (Reuters, 2006). This amounts to a documented rate of 1.6 extinctions per annum, or 0.0032 per cent of the *lower end* of the Greenpeace range.

Moreover, species thought to be extinct have been known to reappear, often in similar niches in other geographical locations. In 2004, a flock of New Zealand storm petrels, thought extinct for 150 years since the last examples were shot to the south of the South Island of New Zealand in the 1850s, was discovered off the coast of the North Island. (While it was decided to take it off the extinct list, it was placed on the critically endangered list, despite any evidence as to its numbers other than those observed.)

This is not an isolated example. A careful watch of the news media reveals at least a steady trickle of species thought extinct which are rediscovered. In 2005, for example, the ivory-billed woodpecker (*Campephilus principalis*), not seen since 1944, was sighted in the same woods in eastern

Arkansas where it was last spotted. It was hailed as the 'ornithological highlight of the millennium', the equivalent of finding a dodo, by scientists. Frank Gill of the Audubon Society likened it to 'finding Elvis' – without the jumpsuit, one presumes (*Weekend Australian*, 30 April–1 May 2005). Millions of dollars were immediately allocated to its preservation, though doubts later emerged as to whether this was a case of mistaken identity, with a vague piece of video footage being the only evidence.

In addition to the storm petrel and ivory-billed woodpecker, several birds thought to have been extinct have been rediscovered, including the rusty-throated wren-babbler (thought extinct for 60 years before being found in 2004 in the Indian Himalayas) and the long-legged warbler (not seen since 1894, but found in the mountains of Fiji in 2003, according to BirdLife International). In Angola, where civil war had resulted in the abandonment of coffee plantations, 18 endemic species not seen and identified by experts for years had been found. These included the orange-breasted bush-shrike (not seen since 1957). BirdLife International still maintained that the overall situation of the world's birds was worsening, claiming more than a fifth faced extinction (Stoddard, 2005). Birds excite greater interest than insects, but insects, too, have been known to re-emerge. In 2004, the jewel beetle, thought extinct for 50 years, was found in the back of a utility truck at Miena in Tasmania (*Mercury*, 23 April 2004). These are by no means isolated examples, even in Australia: the noisy scrub bird (*Atrichornis clamosus*), and Gilbert's potoroo (*Potorous gilbertii*), both described in 1842 and thought to be extinct for 100 years were re-discovered near Albany, Western Australia, the first in 1961 and the second in 1994.

None of this means, or is meant to mean, that conservation of species is not worthy of our concern. But it does call into question the claims of 'virtual' extinction which make for such powerful slogans when applied by groups like Greenpeace. When Greenpeace uses figures of 50 000–100 0000 species becoming extinct every year it is in a world of virtual reality made possible by the dominance of mathematical models, and it would be very hard-pressed to provide evidence of any *actual* extinctions. Budiansky states that the *observed* rate of extinction is but *one* species per year (Budiansky, 1995, p. 165), and this is about the rate according to IUCN-WCU. Indeed, there is reason to suspect that the numbers of documented extinctions are approximately balanced by the numbers of species once thought extinct being rediscovered.

Greenpeace might be engaging in hyperbole to raise concern (and funds). The IUCN Red List lists 12 259 species as threatened world-wide. But the Red List (*Pseudonovibos spiralis* aside) contains only species that have been documented as both existing and being endangered. While there is scientific literature which supports the Greenpeace claim, the problem is the species going extinct in this literature are virtual species: they are presumed to exist and become extinct all on the basis of a mathematical model. This science overcomes the inconvenience of not actually knowing how many species there are (May, 1988) before making a claim that we are in the midst of an enormous extinction 'spasm'. Bob (now Lord) May estimates that science is identifying about an additional net 10 000 species annually, once synonyms identified are subtracted. He too considers that there has been about one certified extinction annually of bird and mammal species over the past century (May, 2005). But, while he admits that we do not know how many species there are, this does not stop him from estimating a *rate* of extinction.

We *should* care intensely about the endangerment and possible extinction of species like tigers and orang-utans, but Greenpeace's virtual extinctions are overwhelmingly likely to be insects, slimes, moulds, and so on, and it is not clear that many people would lose sleep over their passing. There is an argument (in two parts) that they should be concerned. It is one which rests on biodiversity: that in the sheer diversity of species there is both a 'safety in numbers' which makes nature robust against perturbation; and a value for humans in the potential chemicals and drugs which might be found to be of significant medical benefit. The former reason has been thrown into disrepute by the mathematical ecology of May, who showed that complex ecosystems can be highly unstable, although diversity probably provides some advantages in systems recovering from perturbations (May, 1973). And the latter reason, while valid, is often overestimated.

Nevertheless, these are reasonable arguments in favour of preserving biodiversity, but they are not absolutes. It may well be that significantly fewer than the total number of current species provides sufficient resilience, and it cannot be assumed that biodiversity is something of which one can never get enough. Nor is human health necessarily worse off for the extinction of a species – witness the common thought experiment involving the question of whether the world is better off for the existence of the human immunosuppressive virus (HIV). Indeed, this very question arose with the success in eradicating smallpox. We reached the point where the only known, viable smallpox virus was held by the Centers for Disease Control in the USA and a similar repository in Russia. Cowpox exposure provides cross-immunity for humans, so there was little medical value in deciding not to destroy the last remaining genetic material, yet scientists argued against destruction on essentially the same grounds as used to argue for biodiversity.

This chapter will examine the extent to which the politics of the noble cause of conservation of endangered species has corrupted the conduct of the science of conservation biology. We will look first at some examples of the political use of endangered species, and then turn to examine how such political concerns have infused the practice of the science, and how this has been facilitated by the reliance of conservation biology on mathematical models in the form of the species–area equation.

THE POLITICS OF CONSERVATION BIOLOGY

There are several cases of at least dubious science in the area of conservation biology, the most notorious of which involves the production of scientific evidence to strengthen the case for the preservation of old-growth forests in the US Pacific Northwest. Ultimately, in the Pacific Northwest, the northern spotted owl won out over the loggers and the logging companies not just because of the campaigning skills of the radical environmental group Earth First, but because 'science' was produced to strengthen their case.

Demographer Russell Lande was drafted by the campaigners in the Pacific Northwest to produce a scientific paper which could be used to claim that logging would harm the spotted owl (Lande, 1988). Lande knew little of birds, and had apparently never seen a spotted owl, but he applied a mathematical model derived from the effects of pesticide on insect populations to the spotted owl problem. This was consummate virtual science, with mathematical models supplanting observational data, or even conservation biology. Activist Andy Stahl put Lande in touch with scholars who suppled the data, and then helped find peer reviewers willing to write supporting letters (Chase, 1995, pp. 256–7). This was necessary because the only 'science' then available on the spotted owl was an incomplete doctoral dissertation. The Lande paper was created to suit the political campaign and was used together with the notion of precaution to win the day. Whereas it assumed an owl population of 2500 and further assumed that logging old-growth forest would cause its extinction, subsequent research showed the species was far more numerous and, if anything, preferred *regrowth* forest. Regrowth forest provided more prey and more conducive hunting conditions than old-growth forest.

Alston Chase, who recounts this tale of the production on demand of scientific literature to justify the preservation of putative spotted owl habitat from old growth logging, previously chronicled a similar tale of somewhat dubious counting of wolf populations in Yellowstone National Park, where data on animal populations were made to fit the preconceived theories of 'natural ecosystem management' that were ultimately to prove the ruination of Yellowstone (Chase, 1987).

There are two problems with this kind of approach. First, the study of animal populations turns out to be an inexact science, where numerical estimates depend crucially upon all manner of assumptions, which in turn can reflect various values, theoretical dispositions, and even political agendas. Ever since the presence of the snail darter – a small fish – stopped the Tennessee Valley Authority's Tellico dam in its tracks in the 1970s, the Endangered Species Act in the USA and similar measures elsewhere have meant that the opponents of any project have been dealt a trump card any time they can locate an endangered species. This creates what Ian Boale has termed a 'value slope' which then predisposes us towards acceptance of some kind of 'scientific' findings – in both the conduct of science and more consistently in the public discourse which surrounds environmental science.

The second problem is that much 'political ecology' rests upon a problematic foundation and the virtual nature of much of the science surrounding biodiversity and conservation biology provides ample scope for value slopes to take effect. The extensive use of mathematical models in the study of ecosystems provides ample scope for 'science' to be conducted on the basis of models rather than on observational evidence. In the absence of hypotheses which might be falsified by observational data, the extensive use of mathematical models introduces a virtual scientific landscape where species, real and virtual, live and die, and where their utility to noble political causes restricts the scepticism of those who might question the validity of such 'science'.

Endangered species become not just trumps, but face cards in the game of politics, used to create advantage. For example, in Australia, the Tasmanian Greens invoked risks to the endangered orange-bellied parrot in 2003 in opposing a proposed wind farm in the state's northwest. The same bird had earlier been used to advantage by the opponents of the Point Lillias chemical storage facility in Victoria after one individual parrot was seen feeding near the site. Ironically, the cancellation of this project was mourned by the conservation biologists working on the orange-bellied parrot, because the developers had proposed to provide substantial funding for conservation measures, such as fencing areas to protect them from predators, and cancellation of the project ended this prospect (Kellow, 2005). Similarly, in February 2004 English Nature and the Royal Society for the Protection of Birds threatened legal action against the Shell Flats wind farm on the threat it posed to the common scoter (*Melanitta nigra*), a rare little black duck. This is all very well when the species and the threat are well documented, but sometimes the ability of an endangered species to create a trump card presents too great a temptation, and this is when virtuous corruption can occur.

The orange-bellied parrot appeared in another political role in 2006 – one that demonstrated the dangers in invoking endangered species for political purposes, because this time the merest hint of a parrot, together with some mathematical modelling and the precautionary principle were used by the Australian Commonwealth Environment Minister to disallow the construction of a wind farm that was environmentalists' preferred response to climate change, but opposed by residents in a marginal government constituency. The authors of the report into the impact of the Bald Hills wind farm on the orange-bellied parrot produced modelling on the basis that the birds spent time at most of the sites of wind farms in Victoria, despite the fact that the birds had not been recorded at 20 of the 23 wind farm installation sites along the coast of Victoria, even after active searches had been conducted (Biosis, 2006). Only one or two sightings had been made at the other three sites. The authors then assumed that the birds would remain present within a single wind farm location for six months – the longest possible period the migratory species could remain at a winter site, and longer than any bird had been recorded at any site. They also assumed the parrot would make two passes through the Bald Hills site. They did all this to err on the side of overestimation of impact. So while no parrot had been sighted within 50 kilometres of the proposed site, the minister then acted in accordance with the precautionary principle (and an election promise) to block the Bald Hills wind farm on the basis of cumulative impact – compounding the precaution already embedded in the assumptions underlying the modelling.

The political usefulness of the orange-bellied parrot drew attention to other cases in Australia, many of them involving similar use of endangered species by environmental groups. The pebble mound mouse, thought to be near extinction, was used by conservationists to try to stop an iron ore mine at Marandoo in Western Australia, until 21 studies by 60 scientists found the only reason it had been lost was that nobody had been looking for it (Stevens, 2006). Coxens' fig parrot proved particularly useful. With no confirmed sightings since the 1980s, it was reported in the catchment of a proposed dam in Queensland (Roberts, 2006). With almost $1 million having been spent on a recovery programme to save it from extinction, the Parks and Wildlife Service was convinced it was still extant, though an ornithological consultant engaged in the 1990s to search for them (without success) thought the 30 or so unconfirmed sightings to be 'dodgy' and 'politically motivated'.

Similarly, it did not take long for the ivory-billed woodpecker to be harnessed to a political cause – if, indeed, that was not the reason for its 'rediscovery'. In 2006 a US federal judge stopped work on a $320 million Grand Prairie irrigation project on the basis that the ivory-billed woodpecker might have resurfaced, handing a victory to environmental groups that invoked the Endangered Species Act (*Guardian*, 21 July 2006). The case brought against the Army Corps of Engineers by the National Wildlife

Federation and Arkansas Wildlife Federation rested on the project being located 14 miles from where the woodpecker was possibly spotted.

Those who wish to use endangered species legislation as a trump card in environmental politics do not always leave the presence of a suitable species to chance, or even to modelling. A worrying example was the apparent planting by US Federal Fish and Wildlife Department officers of fur from endangered (at least in the USA) Canadian lynx in Wenatchee and Gifford Pinchot National Forests in the Pacific Northwest in 2002. When found out, the officials claimed that they were merely trying to test the reliability of testing methods, by covertly seeing whether the testing laboratories could identify real lynx fur if not told in advance. Critics suspected the samples had been planted in an effort to protect the national forests from logging, mining and recreation. The Executive Director of the Forest Service Employees for Environmental Ethics termed this response 'a witch hunt in search of a false conspiracy' (Wilkinson, 2002). It was, he claimed, 'really about well-intentioned scientists trying to make sure a process works properly but who got caught crosswise by political actors who took what happened and twisted it'. The critics, led by Republican Chair of the House Resources Committee James V. Hansen, were viewed by 'some environmentalists' as leading an attempt to discredit legitimate wildlife research, the conclusions of which sometimes clashed with the interests of mining and logging interests. The problem for the credibility of Forest Service Employees for Environmental Ethics was that their Executive Director was Andy Stahl, the very same person who had arranged the creation of the spotted owl paper. His claim that 'At no time was there any attempt made by the scientists to fabricate a lynx presence' would hardly have been convincing to sceptics.

The prestigious science journal *Nature* editorialized in support of those who had faked evidence of Canadian lynx, labelling the critics a 'lynch mob' (*Nature*, 2002; Dalton, 2002), and was quickly taken to task (Mills, 2002) for supporting this unjustified planting of samples by a researcher involved in this project and another project, the integrity of which was impugned by the fakery (Schwartz *et al.*, 2002). The Forest Service had in 1998 contracted John Weaver who worked for the environmental group the Wildlife Conservation Society. He had reported lynx hair in both Oregon and Washington in areas where nobody expected them. These results were used in a Forest Service application for listing the lynx as an endangered species, but the samples were later found to be from bobcats and coyotes. Further evidence could not be found, so in the 1999 and 2000 survey seasons seven employees sent in samples labelled as wild lynx. While they claimed to be testing the laboratory, they were discovered only because a fellow employee blew the whistle the day before he retired (Strassel, 2002).

It is not just that species that are endangered are used politically, but that the very concept of endangerment is contaminated by other agendas. The most obvious example of this is the extent to which animal rights concerns obtrude into conservation biology, often to the serious detriment of both conservation biology and policies to manage ecosystems and endangered species. This is the case with the international whaling regime, where a regime which originally embodied an agreement among whaling nations about conserving whale stocks for the orderly development of the industry is in danger of collapsing because animal rights arguments have succeeded in extending prohibitions to species which are *not* endangered, triggering exit and talk of exit by whaling parties (Mitchell, 1998). There is a case to be made for the conservation of species like minke whales, but it is not one of species endangerment – it is one based upon the sentience of whales and other qualities which entail just as much 'speciesism' as the anthropocentric views of nature which many environmentalists reject. Similarly, the Yellowstone tragedy owed much to public criticism of practices such as the slaughter of elk – justified on ecological grounds when predator species and indigenous hunting were no longer present, but unjustifiable in the eyes of those concerned with animal welfare.

It is the political blending together of animal welfare and environmental concerns which results in the undue attention accorded to 'charismatic megafauna' in much political ecology, and it has consequences. We might refer to this as the 'Free Willy' problem, after the raising and expenditure of more than $25 million dollars on the ultimately futile attempt to return Keiko the movie star orca – not just to the wild, but to Iceland, where his pod was last sighted, in a romanticized attempt at family reunion. Keiko appeared to crave continued human contact and eventually departed Iceland and turned up in Norway, where he eventually died. Paul G. Irwin, President of the Humane Society of the United States, argued in a press statement that 'By Saving Keiko, We Save Ourselves' (Irwin, 2004), but we might have saved several endangered species (or impoverished humans) for the money spent on this futile attempt to save the life of celebrity megafauna. Even in death, Keiko was exploited for political gain, with Kaare Olerud, of the Norwegian Organisation for the Protection of Nature claiming the carcass posed a toxic threat because it 'could contain' about 0.5 kg of PCB. 'It's a potential threat, nothing has been done to prevent it,' Olerud warned (*ABC News Online*, 11 January 2004).

Such sentimentalism over animals can stand in the way of conservation efforts. It has, for example, been difficult to separate ecological and animal rights issues in the Convention on International Trade in Endangered Species (CITES) on the question of the conservation of the African elephant, where northern herds might be at considerable risk but southern conservation has been so successful that culling has become necessary to prevent the devastation of the landscape as a result of overpopulation. Only grudgingly have the parties to CITES allowed limited trade in elephant products from southern populations, which are far from endangered.

But it is not just with elephants where the major regime dealing with trade in *endangered* species is applied to species whose endangered status is at least problematic. There are, for example, far more tigers than the numbers often quoted would suggest. It is frequently claimed that there are only around 5000–7500 tigers left in the wild and 1000 in zoos. Even if we accept those numbers, the discovery of a tiger in a New York apartment in 2003 (when it attacked its owner) points to a phenomenon which suggests that tiger conservation is *not* about mere numbers. In the United States alone, there are estimated to be as many as 13 000 (and as few as 5000) tigers in private hands (Kapp and Ramsey, 2003). Their owners are forbidden by national laws giving effect to CITES obligations from selling skins or other products, and there has been evidence of illicit trade including (incredibly) meat substitution. Strange as it might seem, there is a shop in Chicago which legally sells lion meat, and there is evidence of tiger meat substitution. Prohibitions on *any* commercial use of tiger products prevent an obvious response to the problem of poaching tigers from wild populations – flooding the world market with legitimate tiger bones and other products, which would drive down prices and lessen the incentive for poaching.

Tigers in the wild are worthy of conservation on all manner of grounds: habitat conservation; preservation of genetic diversity; animal rights concerns. But the argument that tigers as a species are in danger of extinction appears, well, specious, and these other justifications might be weaker when balanced, for example, against the risks wild tigers pose for villagers in places like India where tiger conservation has a somewhat different social meaning than for those living in western cities.

NGOs, of course, use 'charismatic megafauna' as promotional emblems, to draw attention to the problem of biodiversity loss. But this is not without consequences. By far the largest number of species at risk in the various estimates of how many become extinct every year are *not* cute megafauna, but rather less cute species like slimes and moulds. To give an example from Australia, it has been estimated that there are around 309 000 species of vascular plants (including those not yet described), of which slightly more than 22 000 are thought to be in Australia; around 4000 of these are yet to be named and described, and 1168 are regarded as rare or threatened. But there are thought to be almost 300 000 species of non-vascular plants in Australia, of which 187 000 are yet to be named and described (Burgman, 2002). Burgman has argued that the systems for listing threatened species are responsive to the subjective preferences of scientists, are largely unresponsive to true threats, are self-perpetuating and accentuate bias with each iteration. He argues that this focus on the vascular and the vertebrate might actually exacerbate extinction in the neglected group, but also of concern is the virtual nature of many of these species.

If a species has not been described, how can we consider it a species? This is not just a philosophical question (akin to questions about trees falling in forests making a noise or otherwise), but a serious scientific issue. The test of a species is that its members can breed only with other members of the species and produce fertile progeny; species which have not been described can hardly be said to have passed the test of being distinct species, rather than different varieties or sub-species. Species originating from the calculations of a mathematical model can at best be virtual species, but there have been cases where there are doubts over whether a species is at risk or just a subspecies or even a different population. One such case was the northern spotted owl, which was not thought to be genetically distinct from the southern spotted owl; and the red-tailed black cockatoo which stopped construction of a pulp mill in South Australia because a population of 1000 was nearby – despite the fact that its numbers were abundant in other regions. (There is an argument sometimes heard that diversity *within* species also provides resilience in an evolutionary sense.)

As we shall see below, just whether the world is a better or worse place for the extinction of a few thousand fungi or a single megafauna species is moot, and the whole question of extinctions and biodiversity is problematic, at best. What does appear to be significant from this, however, is that ecological science frequently embodies the preferences and values of the scientists practising it. This is an important point, because most scientists take umbrage at any suggestion that personal biases inform their work, but they fail to understand that this is both to a greater or lesser extent unavoidable and need not necessarily be self-serving. It is unusual, however, to find a complete dissonance between the research of scientists and their political affiliations – rather, the opposite is usually the case. For example, Lord May's work on modelling the relationship between complexity and stability in model ecosystems (May, 1973) certainly does not clash with his role as a trustee of WWF and his quite overt support for environmental issues.

May's most significant contribution to ecology was to show that complexity was no guarantee of stability, so that even complex ecosystems should be regarded as vulnerable to disturbance. He has been quoted as saying that his science led him to those beliefs (*Guardian*, 30 October 1999), and he may well be correct, but his work (as he moved from theoretical physics to ecology in the early 1970s) did lend much support to environmentalism because it drew attention to the fragility of ecosystems, and his values might have spurred at least his curiosity. But, significantly, he also typified the rise of virtual science in ecology, because he was by training a theoretical physicist, not a field biologist.

The very science of ecology, to which Lord May moved, certainly reflects social and political beliefs. Ecology is full of terms like 'natural enemies' which were first used metaphorically, but are now frequently used non-problematically and in different contexts to that of their first usage. Chew and Laubichler (2003) have concluded that many, if not most, ecological concepts reflect familiar cultural experience. They note that the discipline is replete with value-laden terms such as 'alien', 'colonize', 'community', 'competition', 'contest', 'disturbance', 'efficiency', 'enemy', 'invasive', 'native', 'stability' and 'territory'. We can add others to their analysis, including 'collapse' and 'threshold', a term borrowed from physics and now pervasive in ecology, and while it is sometimes used with justification, it is often invoked as a threat, a point which once crossed, can never be returned to. This language reflects emotional connotations which are culturally biased and which draw scientists and the public alike towards views that may be at odds with fundamental biological principles. There are many dangers inherent in words like 'natural' and 'unnatural' and the danger of teleology is omnipresent, as the notion of a climax community can suggest a purpose, or a natural or divine design at work.

One danger relates to the second problem foreshadowed above: the assumption that nature is balanced and harmonious, and that this 'sacred balance' (Suzuki and McConnell, 1997) depends upon maintaining high levels of biodiversity. Those espousing this view are in contradiction of May's work, which suggested instability could result even from highly complex systems. Philip Stott (1998) points out that, despite the cogent critique of the idea of stability as the norm in ecology from about 1910, its practitioners continue to speak of climaxes, optima, balance, harmony, equilibria, stability, and so on – and to focus on the 'exotic other' of rainforests and the giant panda (noting that WWF eschews the smallpox virus and the rat as symbols of biodiversity in favour of this charismatic megafauna). He argues that the language which depicts fire, drought, seasonality and cold as 'ecological stresses' is possible only if we maintain a misplaced norm of stability.

POLITICAL ECOLOGY AND 'POETRY'

The second problem identified above lies with the extent to which various theoretical assumptions underpin modern political ecology. These relate to the prevailing belief that nature exists in some delicate and harmonious balance, so that any anthropogenic interference such as actions causing species extinction might trigger catastrophic ecological collapse. But they also relate to the virtual nature of many of the claims for widespread species extinction which is putatively occurring.

Stephen Budiansky has pointed out how much of modern 'political ecology' (that which forms the discourse of environmental activists) is good poetry, but bad science (Budiansky, 1995). Ecologists once thought that nature left free of human interference would eventually reach the steady state of the climax community, but over the past 30 years the idea of adaptation to disturbance has replaced the concept of the climax community among most ecological scientists. It is a point of some interest that in the popular imagination, the stability of the climax community is probably still the dominant 'myth of nature', sustained by constant repetition by political ecologists, and like 'sustained yield', the progenitor of 'sustainable development' (which emerged in a social context of great uncertainty in Germany), no doubt offering the reassurance of stability in uncertain and rapidly changing times. Similarly, 'climate change' suggests that the climate doesn't usually change, which geological science tells us is poppycock.

An ecological science in which perturbation, turbulence, disturbance, succession and flux are the norm creates what appear to be insurmountable problems for ecocentric philosophical positions (see Scoones, 1999). If nature is constantly in flux, how can we read a prescription from it that any particular state should be preserved? If we should leave it to nature to find its own balance, it is likely instead to demonstrate considerable change, and if we wish to preserve any environment we will have to manage it in line with *human* values. While we are not reduced to seeing nature in purely utilitarian terms, this does place the emphasis back on human choice – in Botkin's terms (Botkin, 1990), we must choose among the 'discordant harmonies' of nature those elements we wish to retain. We must reject nature as providing a source of norms which guide how we must live and accept instead that we are part of a living, changing system; we can choose to accept, use, or control elements to make for a habitable existence, both singly and individually. But we cannot leave nature alone to look after itself and expect that we will necessarily approve of the result, since it might not be conducive to either maximizing biodiversity or conserving endangered species, if we prize those values.

An emphasis on disturbance and chaos also suggests we need be cautious about assuming we can manage resources at sustained yield, of course, and this is the basis for the emergence of the 'precautionary principle' – although this is frequently little more than a slogan with an infinite number of meanings. And while Donald Worster (1993) dismisses sustainability as a sloganeering approach to environmental problems, his solution lies in the direction of another slogan: biodiversity preservation. He argues that we must make our first priority the strict preservation of the billion-year heritage of evolution of plant and animal life, and thus preserve all the species, subspecies, varieties, communities and ecosystems that we possibly can. We cannot stop every extinction, argues Worster, but we should avoid adding to the tally.

But even 'biodiversity' is frequently used as a slogan, and there are dangers in this, especially with the unquestioned belief that one simply cannot have enough of it. One can search long and hard for critical discussion of *how much* biodiversity we should seek. There is an unquestioning belief that more is both always better and never sufficient, despite there being doubts as to whether the supposed benefits of diversity, such as ecosystem stability (as May pointed out), are real. This is so not just because of the decline in acceptance of the notion of the climax community, but because there is evidence of resilience in simple systems and fragility in diverse systems (Budiansky, 1995, pp. 97–9).

Slogans sometimes make for good politics, but they are a dangerous foundation for science – or policy. A management plan for a national park in Germany was once saved from the efforts of environmental groups to write into it a requirement to 'maximize biodiversity' only at the 11th hour, when ecologists in the parks agency realized the alpine ecosystem had low natural biodiversity (Haber, 1993, p. 39). Politically, it is often supposed that conservation biology and the preservation of biodiversity are two sides of the same coin, and that human management decisions can be 'read' from the available science in some way. Yet this cannot be taken for granted, as a single example shows.

On Cape York Peninsula in the Australian state of Queensland, an area of wet sclerophyll forest is being 'invaded' (note the bellicose language) by rainforest (Harrington and Sanderson, 1994). The wet sclerophyll is habitat to several endangered species. What management decision should we take? Rainforest ecologists might argue that rainforest is naturally more biodiverse than wet sclerophyll (and, presumably, more stable), and that therefore the invasion should be allowed to proceed. But if we care about species extinction, should we not intervene? And what difference does it make that both the origins of the wet sclerophyll and the resurgence of rainforest are anthropogenic – the former in the fire activity at the hands of indigenous people, and the latter at the termination of that practice with the development of the cattle industry?

Should we privilege the wet sclerophyll (which is an artefact of pre-industrial human fire activity) over the rainforest which is advantaged by the fire suppression activity of modern society? Rainforest ecologists tend to argue for rainforest, citing its greater inherent biodiversity, but conservation biologists would argue that acting to preserve the wet sclerophyll enhances biodiversity on a global scale. As this example shows, all environmental management requires the exercise of human judgement about whether or not to intervene. A decision not to act is just as much a decision as one to do so. And just as with other areas of public policy, the notion of a totally scientific public policy is a myth (Formaini, 1990).

Moreover, there might be dangers in attempting to follow a scientifically reductionist path of public policy-making, both with conservation biology and with other environmental problems. There is at least a need to avoid assumptions that slogans such as 'intermediate technology' will deliver the right results. The tragedy of arsenic poisoning from tube wells in Bangladesh, which have lowered the water table and mobilized naturally-occurring arsenic in geological formations, serves as an appropriate warning. There are numerous examples of the dangers of reductionism, perhaps none more stark than the introduction of ozone-depleting chlorofluorocarbons (CFCs) in 1928 as a safe substitute for ammonia refrigerants. Similarly, we need to be careful about translating risk management decisions from developed to developing nations: Peru following US EPA assessments in deciding not to chlorinate drinking water caused thousands of deaths in the South American cholera epidemic of the 1990s thanks to policy-makers ignoring the differing socio-economic contexts (Anderson, 1991).

It is within this context that some of the recent controversies over species extinction must be placed. This science is now highly politicized, as evidenced by the Greenpeace claims of extinction rates of 50 000–100 000 species annually in its advertising, soliciting funds. The number seems fantastic, but rests upon the mathematical modelling of a supposed species–area relationship developed by Harvard biologist Edward O. Wilson and popularized by Norman Myers. Both are activists and scientists. Regardless, these are *virtual* extinctions, and they lie at the heart of part of the heated exchanges between ecological apostate Bjorn Lomborg and Wilson, Myers and other 'political ecologists' such as Paul Ehrlich, and others still who were quick to mobilize to counter the political impact of Lomborg's book *The Skeptical Environmentalist*, in which he took issue (*inter alia*) with the virtual extinctions resulting from species–area models (see Chapter 4). Two reviewers, Jeff Harvey and Stuart Pimm, likened Lomborg to a holocaust denier because he challenged them to 'name one', ignoring the obvious difference that it is possible in practice to name nearly every holocaust victim if one consults the records whereas it is not possible to name virtual species. But it requires courage to raise issues when one's opponents use analogies like 'holocaust denier'.

Chase has explored the origins of ecosystems science, especially the extent to which it borrowed from physics a model that explained energy flows through a system, with nature operating much like a thermostat, so that ecosystems were seen as self-regulating and tending towards equilibrium. 'An ecosystem could be pictured in the form of a model like an electrical circuit, whereby energy, derived from the sun, was transferred by means of chemical processes through soil and grass up the food chain, and then, by decomposition, through the cycle again' (Chase, 1987, p. 313). Ecology lacked a scientifically respectable method for studying life, and the ecosystem approach provided scientific respectability by supplying ecologists with mathematical tools developed by physicists. It gave them access to the laws of thermodynamics and mathematics. (May, a physicist who became a leading ecologist, typifies the discipline.) Community ecology boomed, but the problem was 'true ecosystems . . . were hard to find. In fact, as even Tansley had acknowledged, they did not exist!' (Chase, 1987, p. 315).

An ecosystem (like a community or society) is nothing more than a construction, 'a tool by which scientists artificially separated their subject of study from everything else' (Chase, 1987, p. 315). Ecologists tried to study ponds as examples of ecosystems, but soon found even they were not closed systems, but connected to the water table, and affected by groundwater flows, spring run-off, migrating waterfowl, trace elements dropped in the rain, airborne spores and the sun. When everything was connected to everything else on the globe, it was impossible to isolate an ecosystem to study, and even earth was affected by sunspots, meteor showers and cosmic radiation. To study nature therefore necessitates the use of models and abstractions, and a degree of reductionism, and this provides opportunities for normative factors to intrude.

Ethics became infused into ecological science. Aldo Leopold tried to develop his 'land ethic' by taking, somewhat to an extreme, the obvious point that man was a part of ecosystems. Whereas Tansley had taken this point to suggest that any separation between man and nature was artificial, and therefore that any human actions were just part of the system, Leopold took it to mean that man should develop an ethical system which included soils, water, plants and animals, or (taken together) the land. But Leopold's land ethic, and Deep Ecology and other ecocentric ethical systems which came later, required a science for studying relations between humanity and nature, and there was none. Environmentalists took to the idea of a self-regulating ecosystem like ducks to Walden Pond, but they failed to appreciate that it was the product of mathematics, part of the very post-Enlightenment rationality they were rejecting as they began to turn ecological science into religion, where knowledge rested on the 'almost sensuous intuiting of natural harmonies', as Theodore Rosak put it (Chase, 1987, p. 323), and the balance of nature was granted sacred status.

The progress of ecological science was to diverge from this notion of a sacred balance, as change, perturbation and succession came to be accepted as core concepts. But the increased emphasis on mathematics which lent ecology its scientific gravitas helped steer it towards virtual science rather than experimental science, and it never totally shook off its normative shackles. The emphasis on mathematical models de-emphasized the need for experiment, and the need for field work. As Chase put it, 'It became perhaps too abstract, a discipline attracting deskbound number crunchers more than those who liked to tramp about the woods in wool shirts counting deer scat' (Chase, 1987, p. 322). Community ecologists reflecting critically on their discipline concluded it had too often been content with generalized mathematical theory and passively (rather than experimentally) collected observations (Chase, 1987, pp. 322–3).

The shift of environmentalism onto a quasi-religious plane, coupled with the descent of much of ecology into the virtual world of mathematical modelling facilitated the virtuous corruption of virtual science.

THE SPECIES–AREA EQUATION AND THE BASIS OF VIRTUAL SCIENCE

What Crichton overlooks in his tracing of the ancestry of climate science to the SETI project via the alarms over nuclear winter is that a substantial legitimation of the heavy reliance upon mathematical modelling at the relative expense of observational science occurred prior to nuclear winter in ecological science, especially in the area of conservation biology and biodiversity, which have been underpinned by island biogeography theory with its species–area equation. This was founded in the 1960s by E.O. Wilson, later of socio-biology fame, and R.H. MacArthur (MacArthur and Wilson, 1967), and serves as the basis for most of the claims of widespread species extinction.

It is perhaps to overstate the matter to suggest that there is a 'theory' of island biogeography, because it consists, Budiansky (1995, p. 165) argues, of a single equation linking the number of species found on an island with the size of the island. There are various formulations of the equation, but that used by Wilson is $S = CA^z$, where S represents the number of species, A represents the area, and C and z are constants selected arbitrarily to get the model to account for the data. On ocean islands, a good fit with the data can be obtained by using a value for z in the range of 0.2–0.4, which Budiansky notes, means that an island ten times the size has twice as many species. This is a useful equation, though hardly earth-shattering science, and the equation can probably be adapted to any set of data to account for deviant cases, simply by tweaking the constants. It is technically what scientists call an 'empirical formula'. As we shall see later, it is the application of this formula to other circumstances that gives cause for concern.

In fairness, there is a little more to island biogeography theory than just the species–area equation – there is also a matter of distance, or isolation. It holds that both area and distance from the mainland determine the number of species found on an island (the equilibrium number), and that these will affect the rate of extinction of species on the islands and the level of immigration. Islands closer to a continental land mass are more likely to receive immigrants than those more remote. In addition to this 'distance effect', the 'size effect' reflects the species–area rule, so that on smaller islands that probability of extinction is greater than on larger ones and larger islands can hold more species than smaller ones as a result. The play between these two factors can be used to establish how many species an island can hold at equilibrium.

The theory was tested by Wilson in the mangroves off Florida. Small islands of mangroves were surveyed by Wilson and his student Daniel Simberloff, and then fumigated to remove their insect and arthropod communities. The islands were then studied to observe the movement of species to the islands. The experimental design was essentially a simulation of the creation of new islands, and within a year the islands had been recolonized, and had reached equilibrium. The islands closer to the mainland had more species present, as had been predicted. But the theory was then extended to non-island situations, on the assumption that an 'island' could be any area of habitat surrounded by areas unsuitable for the species on the island, including 'islands' of remnant 'natural' vegetation surrounded by human-altered landscapes. Thus, reserves and national parks were fragmented islands inside human-altered landscapes which could lose species as they decayed towards a new equilibrium number of species.

This was a politically convenient argument for environmentalists, because it suggested that there was a need to preserve large natural areas, rather than small fragments of wilderness. But it also served at the basis of claims by Wilson and others that we were in the midst of what he called 'one of the great extinction spasms of geological history' (Wilson, 1991, p. 280). The basis for this statement was not scientific observation of a large number of extinctions, but estimates based upon the species–area equation, applied to the loss of natural vegetation. This was also politically useful, because it suggested impacts from activities such as the felling of forests far beyond the area logged, imperilling, Wilson claimed, a vast array of species within an area of a few square miles – tens of thousands of species, many still unknown to science, and playing an unknown role in ecosystem maintenance, especially in the case of insects, fungi, and other micro-organisms.

There are, of course, all sorts of objections that can be raised against this virtual science. One fundamental problem is that it is based upon the erroneous notion of ecosystem stability, since equilibrium lies at both the base of the theory *and* in the prescriptive concern with ecosystem maintenance. We must also accept that an equation derived inductively from insect data in ocean islands necessarily applies to terrestrial ecology. There have been many studies supporting this finding, though (as we shall see below) these might be a mere artefact of the method employed. Moreover, they have not been established for the millions of species that Wilson acknowledges are still unknown to science: the fungi, algae, bacteria and so on that are so important in ecosystem maintenance. The distribution mechanisms of bacteria are substantially different from those of mammals and birds – and the insects Wilson studied to obtain his scientific qualifications – so why should we assume that the same mathematics should apply? Indeed, many of these species are prima facie much more capable of distributing individuals over a wide area, so why should we believe that they are comparable with, say, mammals? And, of course, micro-organisms are probably on balance more capable of rapid adaptation, so why should we not assume that there is not considerable redundancy in the large numbers of such species, so that (if it *were* the case that some essential function in ecosystem maintenance was performed by some single bacterium and that species alone) we might expect micro-organisms to adapt to exploit available niches?

One of the arguments in favour of preserving biodiversity is indeed the notion that there is adaptive safety in numbers, and it beggars belief to think that the millions of species of micro-organisms are so specialized that there is not substantial redundancy in the system. Wilson's view has little scientific basis, and indeed, reflects his cultural evaluation of nature, which he clearly sees as in harmonious balance and at risk from any perturbation. (We will return to such myths of nature in a Chapter 6.) Wilson, of course, has a political agenda, and regardless of what we might make of the 'science' of species–area calculations, he is responsible (if that is not too ironic a word) for the gross inflation of what might be more reasonable estimates of species loss into the 50 000–100 000 used by Greenpeace and others. As Budiansky notes, Wilson has presented widely varying estimates of the rate of species extinction, becoming more prone to hyperbole the further the medium is from the strictures of tight peer review and the more expansive it allows him to be (an ironic variation on the 'distance effect' and the 'size effect'). In the journal *Science*, together with Paul Ehrlich (Ehrlich and Wilson, 1991), its is only a 'conservative estimate' of 4000 species a year that are lost; in his 1991 book *The Diversity of Life* a 'maximally optimistic' 27 000 species are lost each year; while to journalists in outlets like the *New York Times* the number has grown to 50 000 or even 100 000 (Budiansky, 1995, p. 166).

Budiansky describes this as 'a caricature of mathematical ecology, one that does little justice to the rigorous and sophisticated work that has been done in recent years on population dynamics, ecosystem stability, extinction and predator–prey interactions' (Budiansky, 1995, p. 166). While there is a danger in the discipline of too much research depending on mathematical modelling and too little on scientists 'tramping about the woods in wool shirts counting deer scat' (as Chase put it above), the assumptions of equilibrium which pervade Wilson's work run counter to other mathematical work, such as that of May (who did not seem to make the transition from physics to ecology by way of the woods). But even then, May's rise to prominence in ecology without much experience of wool shirt and deer scat helped legitimate an ecological science conducted at the computer rather than in the field (although, being an Australian, kangaroo scat might have been more appropriate in his case).

Wilson has claimed that the species–area relationship has been widely verified, but a review of over 100 studies found it accounted for only half the variation in species numbers on islands, which is hardly great predictive efficacy. More damning, perhaps, was that the authors of this study concluded that the fact that z values tended to fall in the range 0.2–0.4 was nothing more than a coincidence, and that the entire theory might be nothing more than a sampling phenomenon lacking in predictive power – that the more widely you search, the more species you will find (Connor and McCoy, 1979). Moreover, it is considered to be inappropriate that this relationship is extrapolated backwards, as Wilson does, to suggest that reducing the area of a forest will produce the same rate of species reduction as does its growth (Heywood and Stuart, 1992).

The species–area relationship appears to have abysmal success in producing predictions of species loss. Budiansky (1995, pp. 167–8) gives the example of the clearing of almost 90 per cent of the Atlantic coastal forests over the past 500 years, which would leave us to predict the loss of half of all species. Instead, not one known species has been declared extinct, and several birds and six butterflies thought 20 years ago to have been extinct have been rediscovered. The reason for this appears to be that the patches of habitat left intact – the very patches thought by Wilson to be inadequate – have provided microhabitat havens for many of the most endangered species. Nature appears to be more resilient than Wilson's theory suggests and Daniel Simberloff (Wilson's graduate student in the original Florida field research) has argued that island biography theory has been an unwarranted distraction from the main task of conservation biologists in determining 'what habitats are important and how to maintain them' (Budiansky, 1995, pp. 168–9).

If Wilson's views on biodiversity are so questionable, why is he accorded such credibility? One could say that one reason is his reputation: he is a significant figure in the field, and a full professor at a leading university (Harvard). He has received numerous awards for his work: the National Medal of Science, Crafoord Prize, Tyler Prize for Environmental Achievement, Nierenberg Prize, and two Pulitzer Prizes (in non-fiction). After all, he coined the very term 'biodiversity' in 1986.[1] But many of these awards and recognitions have come about *because* of his contribution to the science of biodiversity. He is eminent *because* of his biodiversity views, rather than his eminence serving to protect him from criticism, although Budiansky reports that scientists are more sceptical in private about the high extinction figures cited by Wilson. He quotes conservation biologist Vernon Heywood as stating in an interview that biologists are cautious about expressing sceptical views publicly because they do not wish to be seen to be 'rocking the boat'. 'This is the fear it might damage "the cause" ' (Budiansky, 1995, p. 263, n. 15).

It is thus the noble cause of environmental protection that has protected Wilson's views from critical scrutiny, despite the consequence (above) that they have (by the confession of his own graduate student) inhibited good conservation. There is no doubting Wilson's commitment to the cause. He stated in 2000:

> Had people taken the alert signals seriously, as intelligent people must, [*The Diversity of Life*] would have set the basis for a new level of discussions on the environment and the current ongoing worldwide biotic holocaust exterminating species at the rate of one every 20 minutes. [26,000 p.a.] People might be working on solutions by now instead of still wallowing in ignorance. The facts are clearly and well laid out. The evidence is presented, the theories and data explained at length, at a reasonable cost in paperback (or free from the public lending library). Eight years later people are still presenting in public flawed paradigms (perhaps deliberately) to excuse their gluttonous behaviour which is crushing the planetary life-support systems. (Wilson, 2006)

Wilson is clearly committed to the cause, but Budiansky suggests he is far from alone, arguing that many (but by no means all) ecologists were 'attracted to the field in the first place out of sympathy with the environmentalist political agenda of saving the planet from the evils of technological society and its "mechanistic" worldview' (Budiansky, 1995, p. 164). He suggests that the relative failure of ecology as a predictive science is a plus for such scholars because it frees them from the rigours of the rest of modern, mathematical science to study nature 'more as an artist than as an engineer'.

Regardless of whether Budiansky is correct about the motivations of those choosing to become ecologists, it is certainly the case that many of the leading lights in the discipline have maintained close links with the political environmental movement. Paul Ehrlich for example, in addition to his more obvious involvement with the Zero Population Growth organization, has been on the Advisory Council of Friends of the Earth since 1970 and a member of the Scientific Advisory Committee of the Sierra Club since 1972. He wrote his first alarmist book *The Population Bomb* at the suggestion of David Brower, then executive director of the Sierra Club and founder of Friends of the Earth, following an article Ehrlich wrote for the *New Scientist* magazine in December, 1967, and it was co-published by the Sierra Club. And as we noted earlier, Robert May is a longstanding trustee of the WWF (Worldwide Fund for Nature, formerly the World Wildlife Fund). Thomas E. Lovejoy, now chief biodiversity adviser to the president of the World Bank was Director of the World Wildlife Fund (US) from 1973 to 1987. And Wilson has served on the Scientific Advisory Committee of the World Wildlife Fund since 1978, and on its Board of Directors, 1984–94 and Executive Committee, 1987–92. He is also on the Board of Directors, The Nature Conservancy (1993–) and Conservation International (1997–).

It is this attachment to the noble cause of environmental protection that motivates these scholars to go beyond their narrow disciplinary expertise. Wilson is an expert on ants, Ehrlich on butterflies. It is their political agenda which leads them to proselytize outside the area of their base disciplinary expertise. And it is both the lack of a firm theoretical base (such as physics possesses) and the consequent reliance on empirical formulae – derived inductively from the data, and 'verified' by fine adjustment so that they account for as many data as possible – that facilitate the debasement of science to the extent that an observed rate of extinction of one species a year is ignored in favour of 'virtual data' which allow the convenient claim of 50 000 or 100 000 species extinctions each year. It is probably true that many more than one species becomes extinct per year, and that many of these have yet to be identified, but we are entitled to ask for better proof than can be provided by the species–area equation, and to ask the obvious question: does this matter? We might well still answer with a resounding 'Yes', but the question should not be taken for granted.

Again, we can accept that there is value in diversity, especially from an evolutionary point of view: greater numbers of species might not guarantee stability, but it is entirely feasible to suppose that greater diversity spreads the risk and provides greater insurance against ecological problems arising from the extinction of a species occupying a crucial position in an ecosystem. But as with any insurance policy, we are entitled to ask about the cost of the premium (Cherfas, 1994). Except, of course, if one shares the metaphysical beliefs of Edward O. Wilson.

Wilson sees himself as a 'secular humanist', but also calls for greater spirituality in the environment movement, but with a sound empirical base. But his spirituality eschews questioning the cost of insurance, even to the extent that he considers that biodiversity should not employ economic arguments in its defence. He once stated:

> I also agree with a lot of other environmentalists that to put the price tag on future products that may come from the wild environment and the ecological services that the wild environment gives us – as impressive as these figures may be – is potentially disastrous. Because they leave the impression that the wild environment, the place we live and to which we're so very well adapted, can be bought and sold on the market. And obviously there's vastly more to environmental consciousness than that. (Branfman, 2000)

This points to Wilson possessing what can be called a radical, morals-based political philosophy which is typical of many environmentalists, and which marks environmentalism aside from liberal discourses where matters are fungible and compromise possible, but (despite his claim to humanism) it also points to something close to aesthetic elitism. Wilson obviously finds beauty in nature, which is all very well, but to privilege this above other human needs as measured by opportunity cost is undoubtedly elitist and inherently anti-humanist.

Wilson has at times been remarkably naïve about the social and political consequences of his thought. His foray into socio-biology, for example, excited widespread criticism from the political Left (and saw one critic pour a jug of water over him!). Wilson was accused of racism and was apparently quite unaware of Marxist theory and taken aback by the vehemence of the criticism of his views on socio-biology. And he also appears to remain unaware of the trap of the naturalistic fallacy: that what is or has been our genetic or natural inheritance cannot serve as the basis for ethical prescriptions for future human behaviour.

The species–area equation is nothing, however, compared with the catastrophism it has largely replaced – concern with overpopulation. This concern was essentially a neo-Malthusian one that, while population could increase exponentially, agricultural production could only increase as an arithmetic function. Thus in Ehrlich's 1967 book *The Population Bomb* he wrote of population growth outstripping resources and millions of people dying of starvation in the 1970s and 1980s. He claimed that India could not possibly feed two hundred million more people by 1980, and not to have met anyone familiar with the situation who thought India could be self-sufficient in food by 1971 – a claim dropped from the 1971 edition of the book as the 'Green Revolution' defied environmental critics, and improved transport and storage systems and a sharp drop in the global fertility rate made a mockery of the looming catastrophe.

Ehrlich was by no means alone among the biologists in embracing neo-Malthusianism. An early Robert May warned of the dangers of exponential growth, ignoring the fact that economics had long ago dealt with the impossibility of unrestrained exponential growth – and the obvious point that few populations (natural or human) display the growth of lilypads which choke the limited pond through exponential growth in so many cautionary ecological tales (May, 1971). Just as Malthus had been proved wrong for all the reasons critics on the Left (especially Marx) and Right of economics had argued, so too did the view of Ehrlich and the Club of Rome quickly begin to unravel, as reality failed to oblige the mathematical models.

In the face of a deluge of criticisms that his predictions had not come to pass, Ehrlich countered that many of the statements critics claimed were 'predictions' were actually *scenarios*. Indeed, in the first edition of *The Population Bomb*, Ehrlich wrote:

> The possibilities are infinite; the single course of events that will be realized is unguessable. We can, however, look at a few possibilities as an aid to our thinking, using a device known as a 'scenario'. Scenarios are hypothetical sequences of events used as an aid in thinking about the future, especially in identifying possible decision points . . . . Remember, these are just possibilities, not predictions. (Ehrlich, 1968, p. 72)

Ehrlich did acknowledge in an interview that some specific predictions he had made around the time *The Population Bomb* was published, had not come to pass. However, as to a number of his fundamental ideas and assertions he maintained that facts and science proved them valid. He also cited in his defence a 1994 'warning to humanity' by world scientists – an interesting use of a political statement organized by the Union of Concerned Scientists as defence of the failure to come to pass of his *past* predictions. It is not their lack of familiarity with history and the social sciences which is the most worrying aspect of these pronouncements by these eminent biological scientists. They were, after all, specialists in butterflies and ants, or (in the case of May) physicists in the process of transmogrifying into theoretical ecologists. What is surprising is that their views on matters that were in the realm of the social sciences were accorded such credibility by journalists and the public – largely because their message resonated with the emergence of environmental consciousness from the period of about 1968 onwards (and, of course, it helped build that consciousness).

What is even more surprising is the extent to which their myths of nature were put forward as science, especially the view that nature consisted of a long-term, harmonious balance. Take for example, this statement by Wilson in *The Diversity of Life* (1991, pp. 205–6): 'The historical circumstance of interest is that the [tropical rain] forests have persisted over broad parts of the continents since their origins as stronghold of the flowering plants 150 million years ago.' The persistence of tropical rainforests is an assumption for which there was little enough strong evidence at the time, but it has since been found to be mythical in nature. Rather than having been tropical rainforest for 150 million years, for example, parts of the Amazonian rainforest appear to be perhaps 1000 years old, and to have grown over a system of raised fields, irrigation canals, fish weirs, settlement mounds, roads and causeways and other anthropogenic features constructed between about 100 BC and AD 1100, first described in the 1960s by geographer William Denevan and studied in detail by archaeologist Clark Erickson from the 1970s (Erickson, 1988).

The conventional wisdom was that the alternating scorching sun and annual inundation in the Amazon rendered the region incapable of sustaining large civilizations, and farming is certainly difficult even with modern agricultural technology. But the past civilization has clearly been underestimated, and used a system of mounds and canals to provide irrigation in the dry season and drainage in the wet season. Yet despite the evidence provided by Erickson and his colleagues, there was resistance to the idea of a once-populous Amazon, with environmentalists pushing the 'pristine myth' and natural scientists 'literally yelling' at him when he gave talks at the Field Museum in Chicago. Worse still for the catastrophists, it appears that the rich black soil ('*terra preta*' to locals, or *terra preta de índio* – anthropogenic dark soils – to soil scientists) regenerates even when mined and might be the result of deliberate human inoculation with a bacterium (though this much is speculative). What is widely accepted, however, is that the soil is anthropogenic in origin.

As is often the case with science, this view was dismissed as 'revisionist' by a senior established scholar, Betty J. Meggers of the Smithsonian Institution. It was slightly worrying that Meggers, relied on outdated sources in defending the old view in relation to findings by Heckengerber *et al.* (2003) supporting Erickson. In arguing that 'Other observers deny the possibility of intensive agriculture in the region' Meggers (2001, p. 304) cited publications from 1977 and 1983, predating the revolution which has followed Erickson's work, which was not triggered until he completed his doctoral dissertation in 1988. But Meggers provided an insight into what motivated her adhering tenaciously to her view in a paper in *Latin American Antiquity*, where she stated that not only was the revisionist assessment in conflict with the evidence (a valid argument, though not one now widely shared), but that it 'provides support for the unconstrained deforestation of the region' (Meggers, 2001). There are other indications that rainforests do not require millennia to evolve: scientists were surprised to learn that a mature rainforest had evolved in a mere 150 years on Green Mountain on the island of Ascension, after the Royal Navy in 1843 deposited some plants as part of a scheme for revitalizing what Charles

Darwin had described as a an island 'entirely destitute of trees' in 1836 (*Independent*, 16 September 2004).

There was even worse news in early 2006 for those who see the Amazon as teeming with biodiversity. A new piece of research found that the Amazon was in fact extremely *low* in biodiversity among bacteria, with *fewer* species present than in deserts because of the higher levels of acidity found there. It is microorganisms that are thought to be the most diverse and abundant group of organisms on earth, but only recently has it been possible to study patterns of their diversity. In other words, the prevailing view of biodiversity in the Amazon has been based upon those species which are less numerous, and while rainforests might be more biodiverse in terms of flowering plants, insects and other more noticeable species, it might be that they are not particularly diverse after all (Fierer and Jackson, 2006). Moreover, there is evidence that rainforests – rather than existing in some timeless harmony – actually require disturbance to retain their productivity (Wardle *et al.*, 2004).

There is much mythology about the Amazon; it is supposedly the 'lungs of the earth' – but mature forests produce no net oxygen. It is a wonderful place, and well worthy of widescale conservation, but it has fallen captive to the mythology of eminent scientists for the very best of motives, since they wish to preserve it. But the anthropogenic dragons from which they wish to save it appear to be equally mythical. Hunter-gatherer society is not incapable of transforming nature – witness the use of fire by Australian Aborigines and Plains Indians in the USA. There is a belief that traditional societies lived in harmony with the natural environment and had no significant impact on it (Knudtson and Suzuki, 1992), but there is ample evidence that even hunter gatherer societies modified their environment with such techniques as the use of fire. A prehistory for Amazonia which includes not only settled agriculture, but settled agriculture with a benign environmental impact is clearly heresy and hard to accept for those who adhere to old myths. There are strong reasons why we should oppose wholesale deforestation of the Amazon, but they are not so feeble that we should attempt to deny evidence which 'provides support' for deforestation.

To hold to the belief that traditional societies had no impact on the natural environment is to engage in virtuous or noble cause corruption. The idea that humanity is somehow better off being misled by scholars because it cannot be trusted to collectively make the 'right' management decisions is profoundly and worryingly anti-democratic. Yet it appears to be sufficiently common in matters relating to environmental protection so as to arouse considerable scepticism about the whole enterprise. Of all the conservation biologists, Paul Ehrlich has perhaps enunciated this need for scientific activism and lack of faith in democratic processes. In a book co-authored with his wife, he wrote: 'Rational scholarly discourse is all very well, but it does not hold sway where controversies affecting public policies are concerned' (Ehrlich and Ehrlich, 1996, p. 104).

CONCLUSION

The context within which environmental science is conducted provides numerous factors which might facilitate its virtuous corruption, especially where the science is largely virtual – that is, relying primarily upon the results of mathematical models, be they simple ones or those which rely upon substantial supercomputing, rather than more concrete prediction and observation.

Alston Chase pointed out that ecology is at base a discipline that is not readily amenable to empirical study. If everything is connected to everything else, we have no hope of studying it, and even when early ecologists sought to find ecosystems they could study they found that nature confounded them: ponds turned out not to be isolated, but connected to underground water systems. Ecology has therefore retreated into the world of mathematics, some of it (for example, May's work) is rigorous; but other research traditions depend too much upon virtual science where the values of the scientist intrude far too much. These are eminent scientists who should be aware of what they are doing. But they might well remain unaware of the extent to which their values corrupt their science, and they are not encouraged by an adoring environmental movement to elevate scepticism and falsifiability to their rightful place in the conduct of science.

Climate science is an area of science where virtual methods are widespread and the political stakes are high, because climate change as a result of the emission of anthropogenic GHGs seems to demand a cessation of fossil fuel combustion and the use of alternative energy sources so dear to environmentalists. As we shall see in the next chapter, the combination of virtual science and a virtuous cause again prove a dangerous mix.

NOTE

1. The term 'biological diversity' was coined by Thomas Lovejoy in 1980, but the neologism 'biodiversity' was apparently suggested by staff at the National Research Council as constituting more effective communication when he was preparing a report for the first American Forum on biological diversity.

3. Climate science as 'post-normal' science

> There is no more common error than to assume that, because prolonged and accurate mathematical calculations have been made, the application of the result to some fact of nature is absolutely certain.
>
> **Alfred North Whitehead**

Climate change is an area of science where models inevitably play an important role. There is little scope, of course, for laboratory science, or experiments to be conducted with the earth's climate – although that was the motivation of some early climate scientists who saw, during the Cold War, the possibility of altering climate and adversely affecting agricultural production in the Soviet Union. Climate models are constructed using historical data and then tested against the same data, before being used to make predictions or projections about the future. Climate scientist Stephen Schneider (2001, p. 341) has suggested that the use of historical situations is 'essential to test the tools used to make future estimates', and he calls this a 'systems version of "falsification" ', which is central to the conduct of science, but argues that there are 'no precise analogues to the time-evolving what-if questions of global change disturbances'.

So climate change science relies upon the operation of General Circulation Models (GCMs), which depend upon the availability of supercomputers to simulate the operation of the climate system. These are very expensive to develop and run, and as a consequence there are only a dozen or so full-scale models globally, and these are by no means independent of each other, borrowing code from each other and modifying it to try to improve upon other models. The grids with which they can deal are also (at several hundred kilometres) much larger than many of the phenomena they model, such as clouds, which means they have to be parameterized – small-scale events must be represented by large-scale variables (Kiehl, 1992).

The models generate their own 'climates' and these simulated climates have to be run until they stabilize and then compared with long-term trends in the earth's climate system. GCMs have become more sophisticated, incorporating particulate aerosol effects and ocean currents and have aligned more closely with past observations. Most of the models used produce a warming climate even with constant carbon dioxide, and it was only about 1996 when models were developed that would produce a stable climate with stable carbon dioxide. Previously 'flux adjustments' had to be made to all models to keep them realistic, and the political agendas of many of the modellers was indicated by the finding that many of them preferred that the climate modelling community should keep discussion of the phenomenon to themselves, fearing that 'climate contrarians' would exploit the issue (Shackley *et al.*, 1999).

Yet for all these advances in sophistication, the models yield very similar results to the very simple calculations conducted by Svante Arrhenius a century ago – warming of between 1–6°C for a doubling of carbon dioxide (Arrhenius, 1896). Arrhenius essentially set the expectations of what 'answer' the models should yield, and this and the borrowing of code between models has ensured a high degree of consensus among different modelling groups. One problem for early modellers was the limited availability of data, with nothing approaching global data available until the 1950s. The data issue is a significant one for climate modelling, because virtually all data in climate science must first be processed by computer models before they can be used in GCMs. The data cannot be understood without computer models, and yet without the data the accuracy of the models cannot be confirmed. Temperature observations, for example, have to be manipulated to purge them of the 'urban heat island effect' – warming through cities becoming increasing sources of heat. Even land-use changes can affect climate: Kalnay and Cal (2003) found that about half the warming of the past 100 years was attributable to land-use changes, more than double the previous estimate. Furthermore, temperature records over the 70 per cent of the earth's surface covered by water depend upon proxy readings of the temperature of surface water recorded in ships' logs.

Most of the data being fed into GCMs is thus a long way removed from raw, observational data, and much of it is patchy and not always consistent. Our knowledge of the history of carbon dioxide in the atmosphere depends upon records where high results from scientists in the nineteenth century have been discarded, and where temperature records over the 70 per cent of the surface of the globe covered by water are represented by proxies in the form of the water temperatures underlying the air, measured by non-standardized methods. Moreover, it is limited to largely 50 years, when earth systems are known to have varied over millennia, sometimes (as the geologists remind us) at frightening rates. Most of the data are local or regional in nature, rather than truly global. Indeed, it is really only the models that are truly global, save for some data sets such as remote sensing of atmospheric temperatures by satellites – but only since 1977.

Renewed interest in the possibility that carbon dioxide might accumulate in the atmosphere came in the 1950s, with the Mauna Loa monitoring programme being established at the suggestion of Roger Revelle, and new computing power in the 1960s allowing greater sophistication in modelling. First, in 1956, Plass aroused new interest in carbon dioxide as a factor in climate change (Plass, 1956). Then, Revelle and Suess (1957) predicted that fossil fuels might soon induce rapid changes in world climate. They wrote that humanity was conducting, unawares, 'a great geophysical experiment' on the Earth's climate. To track this 'experiment', Revelle proposed establishing a monitoring station for atmospheric CO_2 at Mauna Loa, Hawaii, as part of the International Geophysical Year. The Mauna Loa station has subsequently documented a steady annual rise in the atmospheric concentration of CO_2, and the Global Atmospheric Research Program was established.

Scientists could not agree on how anthropogenic influences might impact upon climate, with concerns in the mid-1970s centring on the possibility of rapid global cooling (Gwynne, 1975). But mean global temperatures began to warm, reversing a modest decline since the 1940s, and the task – as the growth in computing power permitted new modelling approaches – was to explain this warming, and the growth in atmospheric CO_2 became the preferred causal factor once more. In fact, the change came suddenly in 1976, with what is known as the Pacific Climate Shift, a discontinuity not predictable by modelling a slow steady rise in GHGs. The World Meteorological Organization (WMO) held the First World Climate Conference in 1979, where a World Climate Program was established to coordinate climate research and data collection. There was widespread disagreement on such influences as solar variability, sunspots, and cloud feedbacks and the chaotic qualities of the climate system ensured there was considerable uncertainty over climate science. An unusually hot, dry summer in the USA, was accompanied by testimony from James Hansen to a Congressional committee that he was '99 percent certain' that the heat was a signal of global warming. Political interest in global warming therefore coincided with the development of more and more sophisticated computer models that might be able to be used to run simulation 'experiments' to understand the way the climate system operated and how human influences might impact upon it (Hart and Victor, 1993).

Climate science has therefore emerged as substantially a model-based science, where the impossibility of conducting traditional laboratory experiments leads to a reliance on 'experimentation' by computer simulation. One obvious problem is, as Nancy Cartwright (1983, p. 153) once put it bluntly, 'A model is a work of fiction'. The problem is compounded with climate change, because the phenomena being modelled are extremely complex and chaotic: climate scientists must deal with complex, non-linear coupled systems. Moreover, they must try to build models that must operate on runs of global data that are very limited in time, when the geophysical time-scale of climate is considered.

An additional problem lies with the fact that climate science is such a vast undertaking that no single scientist, nor any group of scientists can master anything but a small part of it. A study of the historiography of climate science will reveal that there is a considerable effort by a small number of scientists to work in many areas of climate science (names such as Hansen, Wigley, Jones, Santer, and Trenberth – and their research institutes such as UCAR at Colorado and NASA's GISS recur). But, as Spencer Weart (2004) has pointed out, many scientists (including those central to the prevailing paradigm) simply have to assume the results of others are correct, so that the science is inevitably constructed by these actors. This is not to say that climate science is *only* a construction, but that it is *inevitably* partly a construction, and depends upon trust in the work of others. This element of trust allows for the intrusion of biases, because some sources (those with whom we are familiar) are trusted more than others (those whom we don't know or who have different political beliefs).

Much of climate science is therefore comprised of virtual science, and this is so not just with climate modelling, but with attempts to extend our knowledge of the climate system back into history. In this chapter, I examine two instances where lapses in scientific standards have occurred, both of which have had the effect of (conveniently) contributing to the political case for action to mitigate climate change. The first involves the misuse of statistics in the emissions scenarios used to drive climate models, which had the effect of overstating the future risks of climate change. The second involves the total reinterpretation of the climate history of the past millennium (the so-called 'Hockey Stick' research), that was picked up as a prominent message in the Third Assessment Report (TAR) of Working Group I of the Intergovernmental Panel on Climate Change (IPCC). It served as the backdrop to the press conference in Beijing where the Report was released. The value of the Hockey Stick came because it suggested that the present period was much warmer than at any time in the recent past, removing the inconvenience of a warmer medieval period which was previously thought to have occurred in the absence of greenhouse gases, and which had been the view of climate history presented in previous IPCC Reports. I then point to numerous other examples in climate science.

STATISTICS, MODELS AND SCENARIOS

Despite the reliance of climate science on computer models, there are serious questions about the ability of these models to serve as the basis for generating prospective views about what might happen to the climate in future. That they are attempting to deal with enormous complexity is one thing, but they are also attempting to deal with non-linear processes. And it is in this respect that many scientists seriously question their ability to serve as the basis for prediction, as did the retired Director of Research at the Royal Netherlands Meteorological Institute and author of what is regarded as the definitive text on turbulent flow, Hendrik Tennekes, in an essay circulated in January 2006. Tennekes quite bluntly stated that there existed no sound theoretical framework for climate predictability studies.

The reason for this was, as Karl Popper had noted many years earlier, that for scientists to be held accountable for the accuracy of their predictions, they must compute in advance the reliability of their computation, and for complex models this leads to an infinite regress. Computations of forecast skill are much more difficult than the forecasts themselves, and the next level of the computation of the skill of the skill forecast is insurmountable, in Tennekes' view when a complex system such as climate is involved (Tennekes, 1972). Weather forecasters are subject to the cold test of falsifiability in the relatively short term, but climate modellers are making predictions, projections, or whatever over many decades and the lack of accountability is less of a problem for them. So the centrality of modelling to the climate science exercise permits the greater possibility of noble cause corruption, because it is an activity free from the accountability that comes from the potential for falsifiability (at least in the short term).

Tennekes is not alone in harbouring serious doubts about the reliability of climate models to 'project', 'forecast' or 'predict' what rising levels of GHGs might mean for the global climate at some time in the future. For example, Reid Bryson was one of the first scientists to suggest that people could change the climate, albeit by a small amount (Bryson, 2006). When he suggested this in 1968 in a talk to the American Association for the Advancement of Science, he recalls that he was laughed off the platform by people who now say human influence is the *only* thing that can change the climate. Because of what he sees as limitations in the science, Bryson endorsed the 'do little' policy response of the Bush administration.

There is still much that is not known about both the physics of the earth's climate and the ability of GCMs to represent that system. But no matter how good the models, they all depend upon the assumptions about future GHG levels fed into them. Early exercises in exploring what might happen with a doubling of carbon dioxide, or similar, were useful, but these were clearly only 'what if' exercises. The IPCC produced a set of emission scenarios which then allowed an appearance of 'prediction' to be attached to possible futures. While these emission scenarios, too, could only be regarded as 'what if' exercises, they allowed the IPCC to feed them into

GCMs and call the result 'projections'. Journalists and political activists then present them as 'predictions'.

Strictly speaking, they cannot be regarded as projections because they generally involve rates of increase in emissions (and concentrations) that are higher than those currently being measured, and projections would be based upon a projection from current observed trends. James Hansen (2002, p. 437) has noted that the scenarios used by the IPCC have a growth rate in the 1990s that is almost double the *observed* rate of 0.8 per cent per year, and he considers their use is consistent with what he sees as a more general failure by the IPCC to emphasize data over models.

Nevertheless, most responsible scientists refer to projections, although this is a technicality observed more in the breach than the observance by politicians, activists and journalists, who commonly talk about what the future climate 'will be'. The scenarios contained in the IPCC's *Special Report on Emissions Scenarios* (SRES) therefore facilitated a move beyond climate scenarios to something either a little more concrete in the form of the IPCC-approved (but still erroneously labelled) projections of the responsible scientists, or substantially more certain in the form of 'forecasts' for journalists and political actors. The problem was, the SRES scenarios were based upon an error, and one which seriously overstated the seriousness of the climate change problem.

In essence, the problem was this: by using erroneous measures of current GDP per capita in developing countries, the scenarios seriously overestimated the emissions likely to be released as these countries developed to achieve a level of affluence comparable to those of industrialized countries now. The internationally-approved method of making international economic comparisons adopted in the United Nations-approved System of National Accounts is to use a measure known as purchasing power parity (PPP), which like the 'Big Mac' index used by the magazine the *Economist*, includes not just the value of incomes, but the prices of what they can buy. The SRES scenarios used market exchange rates (MER) which convert GDP statistics into a common measure such as US dollars at the prevailing market exchange rates. This error underestimated the current level of economic welfare enjoyed by people in developing countries, and because the scenarios saw this increasing to OECD-levels by 2100, they overstated the size of the journey to be made, and therefore overestimated emissions, and thus concentrations of GHGs. These in turn drove climate model 'projections' to higher temperatures. Even the lowest emissions scenario was therefore in fact a 'high case' scenario.

Politically, nobody was prepared to suggest that the desirable outcome of massive wealth gains by the developing world was unlikely, and this was not the first time this error had been made in an international context – for much the same reasons. The UN Development Program's *Human Development Report* had also used MER statistics to present a view that the gap between rich and poor was widening. This erroneous finding was politically useful because it at once strengthened the claim of the developing world for development assistance and suggested that this undesirable result came in an era of increasing market liberalization. But it also made it impossible to judge which countries had succeeded in improving the lot of their people and which had not, so it made it impossible to learn which policy approaches had succeeded and which had failed.

This error was first noticed by a former Australian Chief Statistician, Ian Castles, who made a presentation on the problems at a symposium on human development organized by the Academy of Social Sciences in Australia in 1999. Also on the programme (and also highly critical of the *Human Development Report*) was David Henderson, who had been Head of the Economics and Statistics Department of the OECD from 1984 until 1992. In relation to the *Human Development Report*, Castles took up his criticisms with the UN and the matter was investigated by a group of Friends of the Chair of the UN Statistical Commission, which upheld Castles' critique. Castles and Henderson later found the same mistake in the SRES and launched a scathing critique of this exercise in both the academic literature and in submissions that were considered within the IPCC's deliberative processes, at the instigation of the Australian government (Castles and Henderson, 2003a, 2003b; Nakicenovic *et al.*, 2003; Grubler *et al.*, 2004).

The mistake in using MER rather than PPP as the basis for the SREC scenarios had the effect of amplifying to the point of incredibility the growth in less developed countries over the next century, and therefore of ramping up both emissions and levels of GHGs, allowing even more frightening 'projections' of mean global temperatures in the IPCC TAR. Whereas the Second Assessment Report (SAR) of the IPCC had 'projected' an increase of between 0.9 to 3.5°C by 2100, TAR raised this to 1.4 to 5.8°C. This result reflected not new science (which accounted for only 4 per cent of the change), but the effect of the new 'storylines' provided by the SRES scenarios (Wigley and Raper, 2002). This figure of 5.8°C was then frequently rounded up to '6 degrees' by the political boosters of the Kyoto Protocol, and dressed up as a forecast, with no indication given of the virtual nature of the number – being generated by inherently uncertain GCMs on the basis of equally uncertain (indeed, erroneous) emissions scenarios.

The Castles and Henderson critique was defended by the SRES authors and led to some interesting exchanges among economists, but it is clear that the substance of the critique was valid – as one would expect, given that the issues had been resolved by statisticians within the larger UN system.

But what was revealing about the IPCC was the extent to which it went to try to discredit the critique by attempting to discredit the 'messengers'. Castles and Henderson possessed considerable gravitas, and their critique earned favourable coverage in important news media outlets such as the *Economist*, and *Time*, clearly concerning the IPCC, which went on the attack, in an extraordinary press statement released in Milan at the 9th Conference of the Parties to the Framework Convention on Climate Change (FCCC) (IPCC, 2003).

The IPCC statement included the fallacious argument that 'the economy does not change by using a different metrics . . ., in the same way that the temperature does not change if you switch from degrees Celsius to Fahrenheit'. This statement would have only been valid if the criticism was about expressing, say, the GDP of China in US dollars or euros using MER, but it was about the process of translating the GDP of different countries on to some comparable base before choosing a scale. It was an embarrassing mistake for an intergovernmental organization now chaired by an economist, Rajendra Pachauri. But even more embarrassing was the fact that the IPCC sought to impugn the character of the two critics, labelling them as 'so called "two independent commentators"'. In the version of the press release provided at Dr Pachauri's press conference in Milan this claim was followed by a false statement now deleted from the version archived by the IPCC: 'Mr Ian Castles is a member of the Lavoisier Group, a group founded in Australia, whose sole mission is to oppose anything that aims to protect the environment'. This repeated a calumny that had been denied by Castles in the Castles and Henderson paper some five months previously: Castles was not a member of this group.

The Chairman of the IPCC at the time the SRES was prepared was Dr Robert Watson, the Chief Scientist at the World Bank, and the World Bank itself had transgressed on the improper use of MER measures of economic product, again for political effect. The President of the Bank, Mr James Wolfensohn, used the erroneous *Human Development Report* in speeches and in his foreword to the Bank's *World Bank Atlas: Measuring Development* in order to claim that 80 per cent of global GDP was owned by 20 per cent of the population, a figure based on an erroneous MER-based analysis. Moreover, he continued to do so even while acknowledging it was wrong to do so. He stated:

> I have been criticised by an Australian statistician [Ian Castles] for not drawing your attention to the fact that dollars have different values in different parts of the world and I shouldn't use 80/20 and that it should be something else. Well, let me acknowledge tonight that he's right, but I'm using 80/20 because it's the only way I know how to express it . . . but just accept for the moment that there is an 80/20 split in the world. (*Australian Financial Review*, 3 August 2001, p. 72)

Mr Wolfensohn continued to use the wrong data to the end of his Presidency. Though wrong, Mr Wolfensohn found the data convenient, as Dr Shaida Badiee, Director of the Bank's Development Data Group, acknowledged in a letter to Ian Castles on 13 January 2005:

> As I am sure you recognize, the preface to the *World Bank Atlas* was drawn from speeches that Mr. Wolfensohn has given on previous occasions and he is quite attached to the illustration of global income inequality based on the '80/20' comparison. There is no intention to mislead, although we all agree that comparisons of welfare between countries are better stated using purchasing power parities.

This was despite the fact that the *System of National Accounts* was published in 1993 by the World Bank, jointly with the United Nations, the International Monetary Fund, the OECD and the Commission of the European Communities. The *Atlas* showed just how significant the error in using MERs was, because it presented 'key indicators of development' from which it could be calculated that in 2003 China's gross national income (GNI) represented 12.5 per cent of the world total if calculated using PPPs, but only 4 per cent of the world total when calculated using MERs.

The Castles and Henderson critique created quite a stir; given the standing of the critics, it could not be dismissed lightly. It was considered by a House of Lords Select Committee on Economic Affairs (2005, p. 52) inquiry which elicited more harmful commentary. Professor Richard Tol, giving evidence into 'Aspects of the Economics of Climate Change' stated that the nomination of experts to Working Groups II and III of the IPCC for the Fourth Assessment Report was a highly politicized process under the SDP–Green German government: 'only people with close connections to the Green Party have been nominated to the IPCC'.

Castles and Henderson were able to claim support in rejection of the use of MER by three Nobel Economics Laureates (Sir Richard Stone, Paul Samuelson and Amartya Sen) and three Distinguished Fellows of the American Economics Association (Irving Kravis, Robert Summer and Alan Heston). They has also claimed explicit support in the controversy from Sir Partha Dasgupta (Cambridge), William Nordhaus (Yale), the UK House of Lords Committee Inquiry into the Economics of Climate Change, and numerous others (Castles, 2005). The IPCC remained unmoved and continued to use these dubious, exaggerated scenarios as the basis for its Fourth Assessment Report, so that all its 'projections' of future climate rest ultimately upon scenarios it knows are based on a basic statistical error in international comparisons of wealth.

Martin Agerup (2003, 2004), president of the Danish Academy for Future Studies and an expert in scenarios, was similarly scathing of the

SRES and its use in climate modelling, while John Reilly of the MIT Joint Program on the Science and Policy of Global Change called the SRES approach an 'insult to science', suspecting that the scenario teams started with an emissions projection, estimated the relationship between emissions and growth, and finally calculated the growth rate needed to achieve the desired emissions projection (Corcoran, 2002). This device, known as 'reverse adaptation', was also suspected to have been used in the scenarios for nuclear winter, and has frequently been used by technocrats in electricity planning (Kellow, 1996) and elsewhere to exercise power (Winner, 1976).

Scenarios can be useful in exploring possibilities, but they bring with them problems. The IPCC scenarios were even criticized by some because they assumed that more oil would be consumed by 2100 than existed in current estimates of reserves (Coghlan, 2003) – reminiscent of the use of scenarios in the nuclear winter saga, when it was suspected that nuclear targeting scenarios were selected not for their realism, but for their utility for the 'science' they drove. Scenarios are properly regarded as 'what if?' exercises, and should *not* be regarded as predictions, forecasts, or even projections. In IPCC modelling exercises, the 'predictions' of the media and environmental groups, and 'projections' of the IPCC, all rested on no firmer foundation than a set of scenarios, and could be no better 'science' than those conjectures. That these scenarios included erroneous assumptions that involved a basic statistical error, and that the IPCC continued to cling tenaciously to this mistake, revealed the extent to which a good cause could limit the conduct of science. But if the 'science' of modelling the future provided one example, that involved in the science of reconstructing the past was even more worrying.

THE HOCKEY STICK CONTROVERSY: A TREE RING CIRCUS

Records of surface air temperature (SAT) are limited historically and geographically. While some records stretch back almost as far as the invention of the thermometer, anything resembling global coverage dates back only 50 years or so – a blink of the eye in geological terms. One solution to this problem is to use proxies, such as tree rings, but these can reflect growing conditions (including rainfall) and not just temperature, and then only in summer, during the growing season. Another is to measure directly temperatures in boreholes and compute historical records of ground surface temperature (GST) that are consistent with the SAT record.

David Deming, of the College of Geosciences at the University of Oklahoma, published a paper in *Science* in 1995 that reported that borehole data indicated that about a one degree Celsius warming had occurred in North America over the past 100 to 150 years. He concluded the paper by noting that 'A cause and effect relationship between anthropogenic activities and climatic warming cannot be demonstrated unambiguously at the present time' (Deming, 1995, p. 1577). Deming reported in 2005 that this paper gave him some credibility among climate change scientists, and one of them (a major figure, Deming stated) let down his guard and sent him an astonishing e-mail that said 'We have to get rid of the Medieval Warm Period' (Deming, 2005).

The Medieval Warm Period (MWP) presented a real problem for those who considered anthropogenic global warming to be an urgent problem, because the MWP was considered by established science to at once be both far warmer than the present and a period not just of benign climate, but of favourable climate which stimulated a great flourishing of humanity. Beginning at around AD 900 and lasting until 1300, it was then followed by a cold period known as the 'Little Ice Age' (LIA) from about AD 1500 to 1850. The MWP (sometimes know as the Medieval Climate Optimum) saw enhanced agricultural yields, decreases in infant mortality and substantial population growth in Europe. Greenland was green, and was colonized by the Vikings, with around 3000 colonists finally dislodged by the LIA. This was the broad sweep of climate history previously accepted by the IPCC.

The climate history of the past millennium was troublesome for global warming theory for three reasons. First, it showed that substantial variability was natural. Second, this variability included a period that was substantially warmer than some modelled science was suggesting might occur as the result of anthropogenic factors over the next generation or two. Finally, all the available historical evidence suggested that, rather than being apocalyptic, this warming was beneficial. Moreover, it was so entrenched in the scientific mainstream that it was included in the early reports of the Intergovernmental Panel on Climate Change (see Figure 3.1). Hence, the MWP was a problem for the view that the world was undergoing rapid anthropogenic warming that was dangerous in its implications, and the Framework Convention on Climate Change had committed parties to mitigate climate change so as to avoid dangerous anthropogenic climate change.

The publication of two papers in 1998 and 1999 was therefore extremely helpful for climate change politics, because they suggested that the MWP and LIA could be ignored from a global perspective because they were confined to Europe, and were only regional in character. These papers were written by Michael Mann, then of the University of Massachusetts, and his collaborators, and used multiproxy methods (using tree-rings and other proxies) to present a new global temperature record. The first extended back to AD 1400 (Mann *et al.*, 1998), and the second took the 'record' back

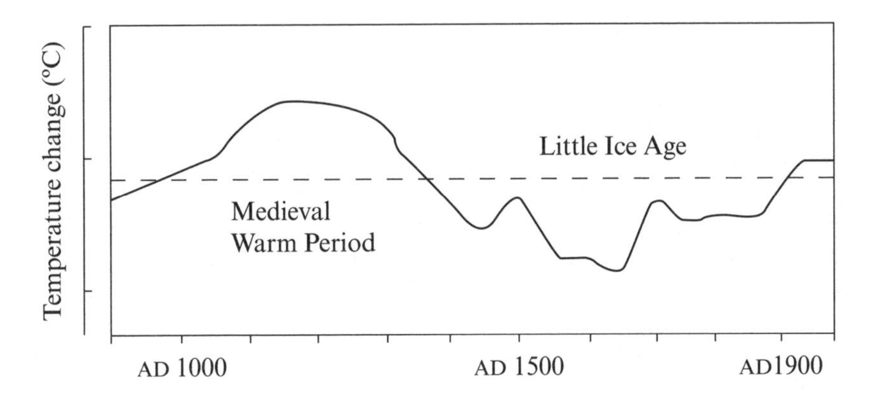

Figure 3.1 *The climate history of the past millennium in the IPCC's first assessment report, 1990*

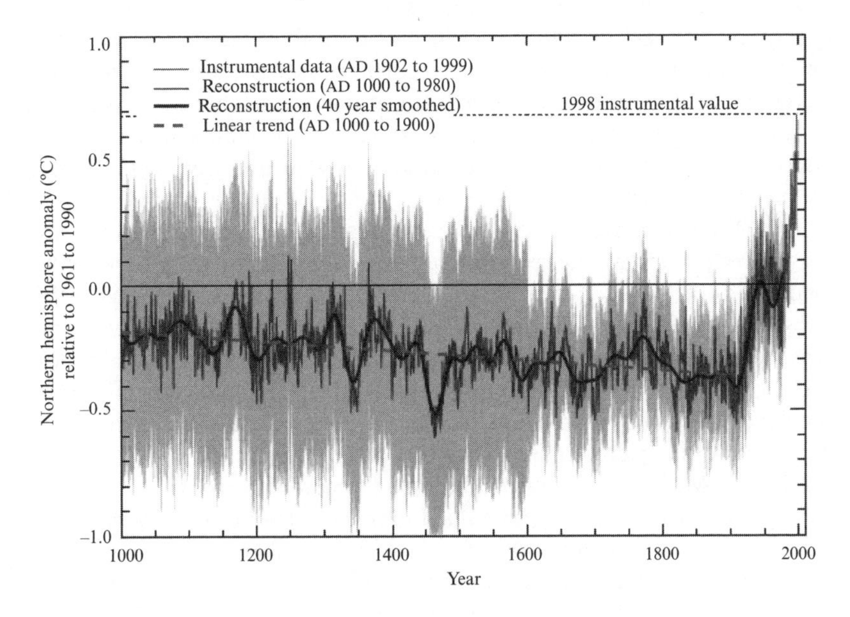

Source: Mann *et al.* (1998).

Figure 3.2 *The 'Hockey Stick' of northern hemisphere temperature anomalies since AD 1400*

to AD 1000 (Mann *et al.*, 1999). They produced graphs with the shape of hockey sticks – relatively flat 'handles' indicating stability from 1000 to about 1900, followed by a rapid upturn in the form of a blade for the twentieth century (see Figure 3.2).

These graphs suggested that the twentieth century was a period of unprecedented warmth, and that this had come about *after* growth in anthropogenic emissions of carbon dioxide from about 1850 as a result of the Industrial Revolution. The University of Massachusetts promptly issued a press release that included in the title the claim that 1998 was the warmest year of the past millennium. This was a claim enthusiastically picked up in the Third Assessment Report of the IPCC (TAR), for which Mann was a lead author of the relevant chapter, in itself a role which raised an interesting conflict of interest. The Hockey Stick featured prominently in TAR and its promotion and allowed IPCC Chairman Dr Robert Watson to repeat the 'warmest of the past millennium' statement to considerable political effect.

What was surprising was not the publication of a couple of papers which challenged the established scientific orthodoxy – that happens all the time – but that these papers were accepted and became the new orthodoxy so quickly and so readily, and it is clear that both this alacrity and readiness and the subsequent defence of the new orthodoxy were inextricably related to the political value of the findings. But it now seems that the enthusiasm to embrace these findings led both the editors and reviewers of the journals concerned and the climate science community to overlook some serious problems with them.

The Mann *et al.*, papers contradicted the picture given by a major analysis of over 6000 borehole records from every continent dating back 20 000 years, with the data for the past 1000 years showing a quite marked MWP substantially warmer than the present followed by a substantial LIA from which the earth was apparently recovering (Huang *et al.*, 1997). Yet this 1997 paper (see Figure 3.3) was all but ignored in TAR, with the only graph included showing borehole data derived from another study showing only a post-1500 curve, conveniently showing a warming trend.

That the Mann *et al.*, view had become the new orthodoxy was apparent by 2003, when Willie Soon and Sallie Baliunas of Harvard University published a paper in *Climate Research* which reviewed 200 existing studies and concluded that there was considerable evidence for both the MWP and LIA. In other words, that the previous orthodoxy was well-supported by evidence (Soon and Baliunas, 2003). The publication of the paper by Soon and Baliunas, which would have been entirely uncontroversial in 1998, provoked the resignation of three of the editorial staff, including Hans von Storch, who had recently been appointed Editor-in-Chief, and had argued unsuccessfully that all papers be submitted to him, which was not the practice at the journal and therefore had not been the procedure with Soon and Baliunas (Kinne, 2003). The consensus now was that the MWP and LIA were just regional phenomena, rather than being global events. In a mere

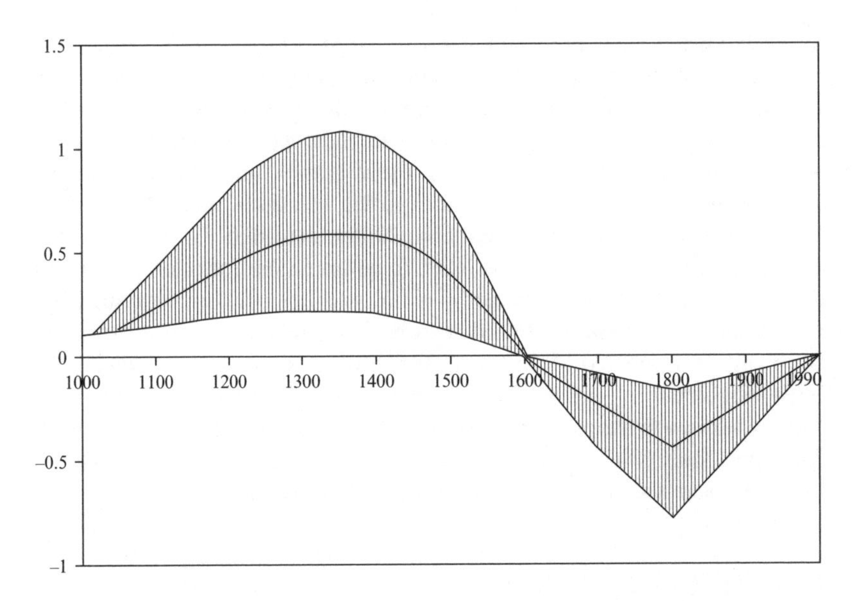

Figure 3.3 The climate history of the past millennium in the borehole analysis by Huang et al. *(1997)*

five years, the established orthodoxy of the climate history of the past thousand had been overturned, and the proponents of the new orthodoxy were prepared to defend it vigorously.

A more direct challenge to the Hockey Stick was to follow. Retired mining executive Steven McIntyre and academic economist Ross McKitrick published a paper in the interdisciplinary journal *Energy and Environment* in which they conducted an audit of the Mann *et al.* (1998) paper (McIntyre and McKitrick, 2003). One of their motivations for writing this paper was a knowledge (particularly on the part of McIntyre) of the need for transparency and auditing of data in the mining industry, where millions of dollars are at stake, and investors in the 1990s had been burned by an elaborate fraud in the case of the Canadian gold miner Bre-X (when ore samples were apparently 'salted' with gold). Fired also by a scepticism about climate change policy, they thought it reasonable that if billions were to be invested in climate change, we should be as certain as possible about the science upon which policy was being based. The Hockey Stick was the most prominent piece of science in TAR, so they decided to subject it to audit.

The initial defence of the Hockey Stick drew attention to the lack of climate science credentials of McIntyre and McKitrick and the fact that the journal in which they published was not comparable in quality to those that had published the Hockey Stick – *Nature* and *Geophysical Research Letters*. But McIntyre and McKitrick were to prove formidable opponents, and to most neutral observers took the points in their dispute with Mann *et al.*, forcing them to publish a corrigendum in *Nature*, and to start a website (realclimate.com) to try to win the public relations war. Review by a National Academy of Science panel in 2006 was to find the reconstruction of temperatures back beyond AD 1600 to be unreliable (effectively restoring the MWP and LIA) and review by a statistical panel commissioned by a Congressional committee found Mann *et al.*, to have misused statistics. Both panels supported McIntyre and McKitrick's criticisms of the Hockey Stick. Significantly, other scientists prominent in IPCC TAR, including Cubasch and von Storch, were also to agree there were significant problems with the Hockey Stick – von Storch's credentials rendered impeccable by his principled resignation over the Soon and Baliunas paper.

MATTERS IN DISPUTE

Mann *et al.*, used a 'multiproxy' method, but the most numerous and most influential proxies in their data set were tree-ring chronologies. The science of dendrochronology used alone is regarded as unreliable, because the size of tree rings do not measure temperature directly, but general growth conditions during the growing season. The size of an annual growth ring thus might reflect nutrient availability, rainfall, temperature, or attack by disease or insects. Significantly, as they reflect the amount of growth during the growing season, they do no reflect the temperature during the winter, when there might be considerable cooling or warming. In one study, tree-ring data from 727 sites showed enormous variability in tree response to climate-forcing, which, it concluded, is influenced by factors such as taxonomy, ontogeny, ecology and environment (Falcon-Lang, 2005).

The method Mann *et al.*, used was a statistical technique known as principal component analysis. It was used because they were mapping a large sample of proxies to a large sample of temperatures, and this gives rise to a mathematical problem that there are more equations than there are unknowns, and PC analysis is used to reduce the dimensions of the data matrices. It involves replacing a group of series with a weighted average of those series, with the weights selected so that the new vector (or principal component–PC) explains as much as possible of the original series. A matrix of unexplained residuals remains, but these too can be reduced to a PC, so that we can have a series of PCs: PC1, PC2, PC3, and so on, in decreasing order of importance in terms of the pattern it explains in the original data.

Often, a large number of data series can be substituted for by a small number of PCs. Mann, Bradley and Hughes (Mann *et al.*, 1998) in their paper in *Nature* (MBH98) used 16 PCs to represent 1082 temperature series, and used 112 proxies (71 individual records and 31 PCs from 6 regional networks containing a total of 300 underlying series).

Early in 2003, McIntyre was curious as to whether the raw data series also looked like hockey sticks, and asked Mann to provide him with the data set used for MBH98, a request with which Mann eventually complied. McIntyre then found that the PCs used in MBH98 could not be replicated, but in attempting to do so he discovered a number of errors, involving location of labels, use of obsolete editions, and unexplained truncations of available proxy series. McKitrick joined the project and they published a paper showing that when the numerous errors were corrected, the Hockey Stick disappeared (see Figure 3.4).

On 9 April 2003, Dr Mann responded to a request from Steve McIntyre for access to the 112 proxies referred to in MBH98, McIntyre having been only able to locate the 13 datasets used in MBH99. He stated that they were available on an 'anonymous ftp site', for which he had 'forgotten the exact location'. He said he would have a colleague, Dr Scott Rutherford, provide McIntyre with the information. Rutherford replied on 11 April stating that

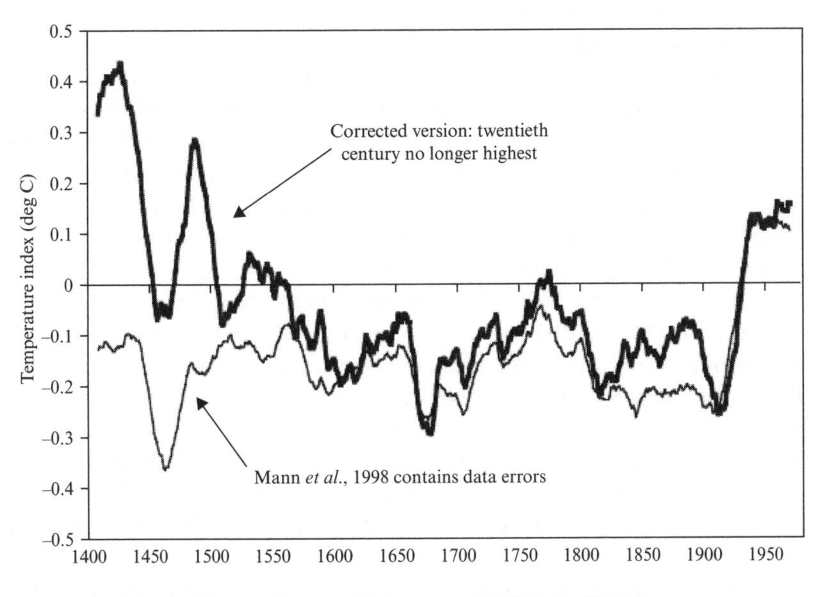

Source: *Geophysical Research Letters*, McIntyre and McKitrick (2005a).

Figure 3.4 McIntyre and McKitrick's corrected Hockey Stick

the proxies 'aren't actually all in one ftp site' but he could get them together in a few days (McIntyre and McKitrick, 2003a). Twelve days later he forwarded them. Mann was later to claim that the data were readily available on his website, and that the data McIntyre and McKitrick used were wrong, presumably having been compiled incorrectly by Rutherford.

Mann did not respond to the first message from McIntyre seeking clarification of problems on September 9, and after another message from McIntyre, Mann replied on 25 September that 'Owing to numerous demands on my time, I will not be able to respond to further inquiries.' Mann was later to complain that the McIntyre and McKitrick paper (McIntyre and McKitrick, 2003) was not sent to him for comment before publication, but it was clearly he who broke off communication. (He launched his initial defence on the website Quark Soup maintained by the freelance journalist David Appell (2003), who was later to write about the dispute for *Scientific American*.) Mann made a crucial error in opening his defence: he criticized McIntyre and McKitrick for using a spreadsheet with 112 columns, 'when in fact the full paleoclimate data series requires 159 to be used.' All previous mentions – including the paper in *Nature* – had been of 112 indicators. This simply encouraged McIntyre and McKitrick to press on in their quest.

Mann's first response was to argue that they had studied the wrong data set, and said that instead of studying the one he had supplied they should have used one in a newly-identified data archive at his university (now the University of Virginia). McIntyre and McKitrick examined this new archive and discovered that it was almost identical to that originally supplied, but it differed in important ways from the description of the data set in the paper as it appeared in *Nature.* They supplied a list of these discrepancies and *Nature* (after an investigation of the issue) required that MBH submit a Corrigendum (Mann *et al.*, 2004). However, MBH were allowed to add a concluding statement that the errors did not affect their results, a conclusion McIntyre and McKitrick disputed.

Mann also objected that McIntyre and McKitrick did not exactly replicate his computations, so they requested from him his computational code to eliminate any trivial differences. Mann then refused steadfastly to reveal his computer code, asserting his intellectual property rights, a stance he maintained until a request from a congressman in 2005 prised it free. It also turned out, McIntyre and McKitrick claimed, that he had spliced together a number of different series in order to deal with segments with missing data in the earliest part of the analysis, again a process not explained in the *Nature* paper.

They then discovered that Mann had undertaken another unusual computational step by standardizing the data. Standardizing is often undertaken to deal with the problem that the data are in different units, so in conventional

PC analysis they are standardized by subtracting the mean of each column and dividing by the standard error, which rescales the data to a mean of zero and a variance of 1. However, this rescaling is not necessary with tree-ring data because the data are pre-scaled before archiving. In addition, however, Mann applied an extraordinary approach, and instead of subtracting the mean of the entire series length, he subtracted the mean of the twentieth-century part of the series and then divided by the standard error of the twentieth-century portion. For most of the series, this made little difference, but for those with an upward trend in the twentieth-century this had an enormous effect, inflating the variance of these series.

Mann's PC algorithm then effectively sought out data sets with a twentieth-century trend and weighted them disproportionately, effectively mining the data sets for hockey sticks. Most of their proxies did not have a twentieth-century growth spurt, but their algorithm gave a weighting to those that did 390 times those that did not. McIntyre and McKitrick ran sequences of 'red noise' random numbers through Mann's algorithm and it produced hockey sticks. In 10 000 repetitions, a conventional PC algorithm almost never yielded a Hockey Stick, but Mann's algorithm yielded a PC1 Hockey Stick 99 per cent of the time.

McIntyre and McKitrick submitted a paper to *Nature* about this flaw, but after an eight-month reviewing process the editors declined to publish it. The editors maintained that it could not be explained in the 500-word limit the editors imposed upon McIntyre and McKitrick (the summary in the preceding two paragraphs alone takes 256 words). One referee stated that he found the material to be quite technical and unlikely to be of interest to the general reader. Instead, MBH were allowed to reveal their non-standard method in an on-line Supplementary Information and then add the unsupported claim that it did not affect the results. The reputations of both MBH and *Nature* were thus protected. (This is explored in more detail below.)

But more was to follow. McIntyre and McKitrick were then able to show that the method gave undue weight to one set of data: that from a group of bristlecone pines at Sheep Mountain in California which exhibited a twentieth-century growth spurt thought to be partly due to carbon dioxide fertilization, and known *not* to be a temperature signal since it did not match nearby temperature records. They had widely been dismissed as not reliable as climate proxies, and – ironically – if they reflected carbon dioxide fertilization, were evidence that rising carbon dioxide levels produced, well, rising carbon dioxide levels. When Mann's algorithm was applied to the North American data after removing the 20 bristlecone pine series there was no Hockey Stick shape to mine the data for, and the result was much the same as that obtained by applying a conventional PC algorithm.

McIntyre and McKitrick reported finding the data for a graph showing precisely this result on Mann's FTP data archive in a folder called 'CENSORED'. Mann had undertaken this analysis himself, and found that – rather than being a global pattern – the Hockey Stick depended crucially upon a set of proxies from a single site regarded as unreliable as proxies of climate. This result had not been reported in the paper in *Nature.*

Further, McIntyre and McKitrick examined the analysis for statistical significance and found that the most controversial pre-1450 segment of the Hockey Stick did not achieve statistical significance. They also found that MBH had used a Gaspé cedar series from 1404, despite the fact that there was only a single tree up to 1421 and two trees up to 1447. Dendrochronologists do not use site data where only one or two trees are sampled. MBH listed the start date as 1400, filling in the first four years by extrapolation, and masking their unusual step of not starting the series from 1450. Removing the pre-1450 Gaspé cedar had the effect of showing a substantially warmer temperature around 1400, falling rapidly to about 1450. MBH's use of this single Gaspé cedar removed any trace of medieval warming.

McIntyre and McKitrick then wrote up the red noise experiment and other issues and had further papers accepted in *Geophysical Research Letters* (McIntyre and McKitrick, 2005a) where MBH99 had appeared and *Energy and Environment* (McIntyre and McKitrick, 2005b). Mann and his co-authors had clearly understimated McIntyre and McKitrick, who possessed the computational skills and determination to take the Hockey Stick apart. And they were joined by mainstream climate scientists, including von Storch *et al.* (2004a), who published a critique in the prestigious journal *Science* in October 2004 which also found that the MBH method was flawed because it tended to dampen out, and thus obliterate, past temperature changes.

Mann and his supporters responded to the criticisms by, among other things, establishing a website, *RealClimate*[1] hosted by the public relations firm Environmental Media Services, a company associated with Fenton Communications, a long-standing promoter of environmental causes and McIntyre and McKitrick founded their own, 'Climate Audit'. But MBH appeared to be losing the PR contest, with Berkeley physicist Richard A. Muller writing a piece in MIT's *Technology Review* accepting the McIntyre and McKitrick critique that the Hockey Stick was an artefact of poor mathematics (Muller, 2004), as well as prominent critical comments by scientists in the Netherlands and Germany.

Muller began to suspect that the Mann reconstructions were suspect after the publication of the Soon and Baliunas paper, and his scepticism grew as the efforts of McIntyre and McKitrick continued. For him, there was an obvious overall problem with the application of PCA to the proxies, because (as he put it):

> the proxy records must be sampled at the same times and have the same length. The data available to Mann and his colleagues weren't, so they had to be averaged, interpolated, and extrapolated. That required subjective judgments which – unfortunately – could have biased the conclusions. (Muller, 2003)

Muller also suggested the mechanism by which the Hockey Stick error had occurred. He pointed out that independent analysis over time and (where possible) independent data sets were the ultimate test of scientific truth, which is why Nobel Prizes are awarded only after the considerable passage of time. It was unfortunate, Muller considered, 'that many scientists endorsed the Hockey Stick before it could be subjected to the tedious review of time. Ironically, it appears that these scientists skipped the vetting precisely because the results were so important.' (The IPCC, of course, facilitated this by allowing Mann to participate in the process which catapulted the graph to political prominence, while simultaneously advancing his academic career.) Muller admitted himself to wanting the Hockey Stick to be true, but pointed out that 'when a conclusion is attractive, I am tempted to lower my standards, to do shoddy work. But that is not the way to truth. When the conclusions are attractive, we must be extra cautious.'

Also of interest was a paper published in *GRL* in late 2005, which found that none of the multiproxy reconstructions of past climate was robust, and it was not clear how they could be salvaged to become robust (Bürger and Cubasch, 2005). The most robust turned out to be the one furthest from the MBH curve. One author, Ulrich Cubasch (a prominent IPCC scientist), had earlier reported that he and his colleagues had also been unable to reproduce the Hockey Stick and told German television that it had dawned on them that the Canadians were right. Hans von Storch was less polite in the German magazine *Der Spiegel*, in a story where the title summed up the content: 'Der Kurve ist Quatsch'. ('Quatsch' translates as rubbish, hogwash, nonsense.)

THE SIGNIFICANCE OF THE CONTROVERSY

The scientific issues in the Hockey Stick case will continue to be played out, and we will continue to gain a better understanding of past climate. That is the way scientific understanding develops. But the case raises some interesting issues concerning the conduct of science.

At one level, it drew attention to some serious problems with the IPCC process, which allowed a relatively junior scholar (Mann was only an

Assistant Professor at Virginia) to be involved in a process which accorded his own work a prominence that enhanced his career at the same time as advancing a politically useful view of the world. That the IPCC embraced the Hockey Stick and gave it prominence while ignoring, relatively speaking, the borehole evidence presented by Huang *et al.* points to a failing in the IPCC process. In particular, this is because borehole evidence is widely regarded as being more robust paleoclimate data than tree rings as the data are not proxies but a direct thermophysical record of temperature changes occurring at the surface (Lachenbruch and Marshall, 1986). An intergovernmental institution developed to present policy-makers with the best possible science ignored the best possible evidence in order to give policy-makers an urgent reason for action.

One hopes that the IPCC will correct these deficiencies, though there are signs that it is somewhat impervious to criticism, as we can see with the SRES case. (The indications are that the drafts of the Fourth Assessment Report basically abandon the Hockey Stick, but make no acknowledgement of the error in embracing it so warmly in 2001.) But perhaps even more worrying is what the case reveals about the practice of one of the leading journals involved in the case, *Nature*. As we saw above, *Nature* acted to preserve as much as possible its own reputation in responding to the complaint by McIntyre and McKitrick. If did so by deciding its readers would not be interested in the technical details of the case and not publishing evidence of considerable error, to say the least, which *Nature* was complicit in disseminating. But Deming has pointed out that *Nature* also declined to publish the Huang *et al.* paper on boreholes, despite a strong positive recommendation by Deming, who has identified himself as one of the referees for the paper (Deming, 2005). Deming told the editors it would be one of the most important papers they would publish that year, but the editors effectively tried to kill the paper, requiring the authors to revise it twice and then, after a long delay, rejecting it. Such treatment frequently makes research lose its currency, and makes it an ordeal for authors to get their paper published, because rules prohibit simultaneous submission to multiple journals, and the paper had to begin anew at *Geophysical Research Letters*.

The treatment of Huang *et al.* by *Nature* closely parallels the journal's treatment of McIntyre and McKitrick, but *Geophysical Research Letters* has shown more willingness to treat the papers by both sets of authors on their merits, despite the negative implications of their research for climate change politics. *Nature* has taken to publishing papers which support the consensus on the eve of each Conference of the Parties to the FCCC, and this behaviour raises serious questions about the objectivity of the editors.

As we have seen, in January 2004, McIntyre and McKitrick submitted a short article to *Nature* dealing with some of the problems they had identified in MBH98, a paper the journal has published with which a growing number of problems was becoming apparent.[2] Their submission letter admitted that the paper submitted did not fit neatly any of the journal's publication categories of 'Letters', 'Articles' and 'Communications Arising', and stated explicitly that they were open to guidance on editorial format and requested that the submission be considered on its merits. In early March they received a favourable 'revise and resubmit' response from the editors on the basis of two constructive referees' reports, and were asked to add additional words in order to respond to these. (The paper stood at 1910 words at this stage.) They added a paragraph to deal with some issues identified, and resubmitted in late March. *Nature* then asked them to reduce the paper to 800 words, which they did, and resubmitted the shortened version on 9 April.

They then did not hear back from *Nature* until 4 August, with the editors attributing the delay to delays in hearing back from reviewers. This was unusual, because the referees had already seen the earlier version, and it should have taken little time to assess whether the revisions met their concerns. But it turned out that *Nature* had added a *new* reviewer. The 4 August communication was a rejection, on the basis that 'the discussion cannot be condensed into our 500-word/1 figure format.' This was largely on the recommendation of the new reviewer, who found no fault with any of the findings per se, but added that the 'technical issues . . . [were] not necessarily of interest to the wide readership of the Brief Communications section of *Nature*'. Most scholars have their own horror stories of what they consider to be rough editorial treatment, but this case is extraordinary, and represents curious behaviour from a journal the reputation of which was on the line for having published MBH98.

Sceptical climate scientist Pat Michaels recounts a similar tale – that a paper showing that deaths declined in US cities despite warming, contrary to suggestions that warming would push them in the opposite direction, was rejected by *Nature* without even going to referees. Despite this rejection, the paper was subsequently submitted to the *International Journal of Biometeorology*, accepted, and then awarded as the 'paper of the year' by the Climate Section of the Association of American Geographers (Davis *et al.*, 2003).

The Hockey Team members have had an easier run with the journals. In May 2005, Stephen McIntyre was asked by editor Stephen Schneider to referee a paper submitted by Wahl and Amman to the journal *Climatic Change.* The paper sought to validate the Hockey Stick, but McIntyre found numerous problems with it, particularly with the statistics employed – failure to provide cross-validation statistics, referring to tree ring chronologies as 'unstandardized' (when they are all pre-standardized), and so on. But the biggest surprise to McIntyre was that the paper had been accepted for publication without reference back to McIntyre, in line with the journal's editorial policies, and that it had been accepted with two curious points uncorrected. The first was a reference (and reliance on for some important points) to another paper by Wahl and Amman being reviewed for publication in *Geophysical Research Letters*. The paper had been rejected by *GRL* once and resubmitted, but after the new paper had been accepted by *Climatic Change* it was *again* rejected by *GRL*. The *Climatic Change* paper therefore referred to (and depended on) a paper that it stated was under review, but was in fact twice-rejected.[3]

Wahl and Amman also (totally incorrectly) described McIntyre and McKittrick as having presented an alternative climate reconstruction. Not only had they always stated that they were doing no more than auditing the Hockey Stick, but McIntyre pointed out the error in his review, and had told Amman of the error when they met at a conference subsequently. Despite this, the *Climatic Change* paper went to press perpetuating the error (which had first been committed by Mann, who wanted to make something of the shape of the McIntyre and McKitrick 'reconstruction').

The performance of the major journals *Science* and (especially) *Nature* has also been suspect in the key area of satellite remote sensing of the earth's atmosphere, an important area of climate science because this is one of the few sources of truly global data. Moreover, it has shown a warming of only 0.085°C/decade, only a third of what is expected from climate models for the troposphere. The satellite data, which commenced in 1978, in fact showed a slight cooling trend at the time of the IPCC Second Assessment Report and it was not featured in that report. Five years later it showed a slight overall warming and that *was* included in the Third Assessment Report (though neither of these trends was larger than the error term in the measurements). So the satellite data posed problems for the pro-warmers.

A paper appeared in *Nature* in 2004 (Fu *et al.*, 2004) that claimed to resolve the discrepancy, used a technique that the two leading scientists in the area, Roy Spencer and John Christy, had tried and rejected in 1992 (Spencer and Christy, 1992). Spencer and Christy are prominent sceptics concerning the evidence for global warming, since they have seen little evidence of it in the data on which they have made their reputations. One might wonder how the paper passed peer-review at *Nature*, but Spencer revealed that it was not surprising, because *Nature* and *Science* had seemingly stopped sending papers to either him or Christy, despite the fact that they are arguably the best qualified people in the field (Spencer, 2004). Spencer said other scientists had had similar experiences with the two journals, and that 'The journals have a small set of reviewers who are pro-global warming.'

In 2005, *Nature* published a paper suggesting that the Gulf Stream had slowed down by 30 per cent (Bryden *et al.*, 2005). Later research failed to replicate these findings, but (again) *Nature* refused to publish correspondence submitted by Petr Chylek of the Los Alamos National Laboratory, pointing out that the observed change was well within the uncertainty of the measurement, and that the correct conclusion from the presented data was that no statistically significant change had been detected, a simple error both the editors and reviewers should have picked up. *Nature* responded to Chylek that his submission would be of no interest to *Nature*'s readers (communication by Chylek to CCNet, 18 October 2006). Chylek (2007) subsequently published his critique in *Physics Today*, and reported an interesting communication with Bryden that the title of the paper in *Nature* ('Slowing of the Atlantic meridional overturning circulation at 25°N.') had, as submitted, concluded with a question mark, which the editors had insisted be removed before publishing on the eve of the 11th Conference of the Parties to the FCCC in Montreal.

It is not just in the area of climate science that the editorial standards at *Nature* have been called into question when science with convenient findings for particular environmental agendas is concerned. For example, Polacheck (2006) pointed to problems with a paper suggesting a 90 per cent decline in tuna stocks attributable to the controversial practice of long-line fishing. Polacheck pointed to errors that should have been picked up in review, but the findings fitted a political agenda, and *Nature* then refused to accept for publication correspondence pointing this out. He argued that the editors of *Nature* should have been aware there were serious concerns in the relevant scientific community about the reliability of the methods used, as these were clearly acknowledged in the supplementary material accompanying the paper. Polacheck summed up the conduct of the editors in the following terms: 'Based on their response, *Nature* appears to divorce itself from responsibility on the accuracy or general validity of the conclusions of an article once it has been published' (Polacheck, 2006, p. 480).

One problem with the safeguards of peer review in many areas of science is that there is a very small group of relevant experts, who all know each other, and are in a position to defend their view of the science from external challenges. So if one looks at the scientific literature in an area like multi-proxy reconstructions of past climate histories back before an instrumental record is available – the technique at the heart of the 'Hockey Stick' controversy – one finds a recurring list of names: K.R. Briffa, E.R. Cook, T.J. Crowley, J. Esper, P.D. Jones, M.E. Mann, A. Moberg and T.J. Osborn dominate the literature, supplemented by a few others, many of them the graduate students or postdoctoral fellows of the eight most prominent. We might add some less prominent names: G.A. Schmidt, or D.T.G. Shindell, for example. Any editor looking for reviewers for a paper on this subject would likely (quite justifiably) turn to two of those named on this list. But the small size of this list means that there is not a wide range to choose from.

Moreover, most of the names on this list have collaborated with others. Science is at once global, and (through increasing specialization) very intimate. Thus we find that Briffa has published with: Osborn;[4] Cook;[5] and Jones.[6] Cook has published with: Briffa;[7] Esper;[8] and Mann.[9] Crowley has published with Mann.[10] Esper has published with Cook.[11] Jones has published with: Briffa;[12] Mann;[13] Moberg;[14] and Osborn.[15] Mann has published with: Cook;[16] and Jones.[17] Moberg has published with Jones.[18] Edward Wegman and his panel (Wegman *et al.*, 2006) performed a social network analysis of this research community as part of his report to the US House Committee on Energy and Commerce, in an attempt to explain how the peer review process had failed. He found close links including co-authorship among the 42 leading scientists in the area, which he considered compromised the independence of the peer review process.

In response to the survey of multiproxy studies by Soon and Baliunas (2003), Mann mustered virtually the whole of the leading multiproxy community to attack their analysis: Ammann, Bradley, Briffa, Jones, Osborn, Crowley, plus climatologists Trenberth and Wigley for good measure (Mann *et al.*, 2003). More worrying still in the whole episode was the opposition by MBH to the notion that they should open their research up for external scrutiny. MBH refused to provide their source code to M&M, arguing that it was Mann's intellectual property. There was, of course, little prospect that there might be any future commercial application arising from the code which would be compromised by it becoming available in the public domain, and this was clearly a device to try to avoid scrutiny – behaviour the very opposite of the scientific ideal of free availability of data and methods to permit verification by replication. Further, the MBH research had been funded by the US taxpayer, but the National Science Foundation upheld Mann's right not to disclose, having no condition attached to grants requiring reasonable sharing of the data at the base of research (in contrast to, for example, the Australian Research Council).

Many of what has been called the 'Hockey Team' have also been reluctant to disclose data. McIntyre reported on his website Climate Audit on 31 July 2005 that Keith Briffa had finally archived earlier that year 13 tree-ring proxies for a paper published back in 1988 (Briffa *et al.*, 1988). He contrasted this with a paper by Alberto Mangini and others published that week in *Earth and Planetary Science Letters* which saw the data archived immediately.

The climate science community is clearly not strongly committed to the principles of disclosure and transparency. McIntyre had difficulty getting most of the 'Hockey Team' to disclose data. He reported in January 2005 that he had had over 25 e-mails with Crowley, who finally said he had 'misplaced' his original data, although he did finally locate some transformed smoothed versions and sent them to McIntyre.[19] He added that he had had a similar experience with Phil Jones, whose data set forms the basis of the temperature readings upon which the view that global temperatures are increasing has been formed. This data, as noted earlier, is subject to manipulation in order to be useful, and the way in which it is manipulated is therefore crucial. McIntyre asked for the data set Jones used in a study that Jones had conducted rebutting the view that this data was contaminated by the 'urban heat island' effect; Jones responded that the data was on a diskette somewhere, but he wouldn't be able to find it. Jones had dismissed a similar request from a sceptical blogger, Warwick Hughes, with a statement almost unbelievable coming from a scientist: 'We have 25 or so years invested in the work. Why should I make the data available to you, when your aim is to try to find something wrong with it.'[20] Hans von Storch was later to tell the National Academy of Sciences panel that he found this statement so incredible that he had sought – and, alas, obtained – verification of its accuracy from Jones.

These 'my dog ate my homework' stories appear to be attempts to defend non-disclosure, but there have been more brazen defences. For example, Ross Gelbspan reports an instance where prominent climate sceptic Pat Michaels requested some model output data from the UK Hadley Centre's climate model as part of his activities as a reviewer for the IPCC. Prominent climate scientist Tom Wigley was quoted as defending the non-provision of data, claiming that it was unnecessary to have original raw data to review a scientific document and stating that he knew of no case in which data had been required by or provided to a referee (Gelbspan, 1998, p. 212). Further, he argued that the data had been generated using the funds provided by UK taxpayers, so US scientists had no right to demand it, and (finally) the data belonged to the individual scientists, not the IPCC. All this speaks of a lack of commitment to norms of transparency and accountability in climate science.

McIntyre, understandably, remarked that Mann's approach to disclosure was perhaps better than the norm, but that approach also included blocking McIntyre's IP address (and that of his neighbours using the same Internet Service Provider) from accessing the Mann ftp site once battle was joined – a somewhat churlish gesture, and ultimately a futile one, since McIntyre was still able to access the site from another IP address.[21] McIntyre also accused Mann of effectively shredding evidence by deleting files. Between 29 October and 8 November 2003 the files 'pcproxy.txt' and 'pc.proxy.mat' were deleted from Mann's University of Virginia FTP site. McIntyre also claimed that after they released the first part of their reply to Mann's criticism of their audit on 11 November 2003, a file that had been available from his old FTP site at the University of Massachusetts, but not on his University of Virginia site, was removed.

PROBLEMS WITH CLIMATE SCIENCE

German climate scientist Hans von Storch (together with two social scientists) has called for a change in the culture of climate science, which they see as having been corrupted by the noble cause at stake:

> The concern for the 'good' and the 'just' case of avoiding further dangerous human interference with the climate system has created a peculiar self-censorship among many climate scientists. Judgments of solid scientific findings are often not made with respect to their immanent quality but on the basis of their alleged or real potential as a weapon by 'skeptics' in a struggle for dominance in public and policy discourse. (Von Storch *et al.*, 2004)

Von Storch *et al.*, suggest that the reason for this outcome is a mix of high policy relevance with high stakes and a high level of uncertainty in the science, a mix exacerbated by the involvement of social movements, a blurring of the distinction between scientists and advocates, intense lobbying by interest groups and their 'indentured scientists' and the emergence of individual scientists as public and media figures.

Another critical essay by von Storch and Stehr was published in *Der Spiegel* on 24 January 2005 (reprinted in translation as 'A climate of staged angst' on the website Prometheus on 7 February 2005).[22] They argued that the mechanisms for correction within science had failed. They stated:

> Within the sciences, openly expressed doubts about the current evidence for climatic catastrophe are often seen as inconvenient, because they damage the 'good cause,' particularly since they could be 'misused by skeptics.' The incremental dramatization comes to be accepted, while any correction of the exaggeration is regarded as dangerous, because it is politically inopportune. Doubts are not made public: rather, people are led to believe in a solid edifice of knowledge that needs only to be completed at the outer edges.

Detailing the response to their paper exposing the methodological weaknesses of the Hockey Stick, they reported that 'Prominent representatives of climate research . . . did not respond by taking issue with the facts. Instead, they worried that the noble cause of protecting the climate might have been done harm.'

Richard Muller (2004) has also put his finger on the nature of noble cause corruption (and the need to guard against it). It does not necessarily involve the deliberate fabrication of evidence – though we know that occurs from time to time for both base *and* noble motives. The danger is that we are much more likely to be suspicious of the influence of economic interests in contributing to a value slope which might be more encouraging for some kinds of results than others. That said, much of the discourse against industry and industry-funded science contains a naïve and populist anti-capitalism which assumes that economic interests will always be disadvantaged by regulation in the name of environmental protection or some other cause. But this is fallacious reasoning; capitalism divides on regulatory issues, because regulation creates winners and losers within business. To use an apposite example, policies requiring decarbonization in an attempt to mitigate climate change disadvantage coal interests, and favour gas interests (mostly the multinational oil companies), the nuclear industry and the renewables sector.

But Muller also points to the circumstances under which the dangers of noble cause corruption are greater: where there is a requirement (and thus an opportunity) for the substantial manipulation of data. It is this room for the possible intrusion of subjective factors into the assumptions which must be made to make data usable, and the use of complex models which are impenetrable to the outside observer which facilitates noble cause corruption. We might, in short, hypothesize that the greater the distance from pure observational data and the greater the use of complex models, the more scope there is for noble cause corruption. The possibility for subjective factors to affect scientific research is common enough in experimental science, and it is even greater in the virtual world of mathematical modelling.

Climate science often seems to prefer model results to observational data. (We noted earlier that even 'warmer' James Hansen levels this criticism at the IPCC.) To give but one result, Chuine *et al.* (2004) presented in *Nature* a mathematical model to estimate April–August temperatures for Dijon in Burgundy for the years 1370–2003 on the basis of the harvest dates of certain grapes. Using the model and historical records of grape harvests, they concluded that 2003 was by far the warmest year of the period 1370–2003. They used the period 1960–1989 for a benchmark, and found the model-estimated temperature for 2003 was 5.86°C higher. But, when compared with the *observational* record, the model overestimated the warming of 2003 by 2.36°C, and suppressed the warmest years in the instrumental record prior to 2003 (1952, 1947 and 1945) so that they appeared nearly average (Keenan, 2007). Or, to give a more important (and policy-relevant example), the Executive Summary of the Report *Temperature Trends in the Lower Atmosphere – Understanding and Reconciling Differences* revealed a discrepancy between temperature observations in the tropical lower troposphere and those values expected from climate models. The report said of the discrepancy:

> It may arise from errors that are common to all models, from errors in the observational data sets, or from a combination of these factors. The second explanation is favoured, but the issue is still open. (Wigley, 2006, p. 11)

This preference for model results and the substantial data manipulation they allow makes even more important the adherence to quality control standards in climate science, yet it appears that these norms are seriously underdeveloped. Reflecting on the Hockey Stick controversy, Roger Pielke Jr (2005b) has suggested that one of the problems is that climate science does not have a well-established tradition of data archiving and transparency. Neither do many of the journals have a tradition of double-blind refereeing; a colleague stated to me that he had never known the identity of authors to be masked in any of his publication or refereeing experience. Earth scientists sometimes justify this on the basis that it allows authors and referees to interact, to discuss interesting points that might arise from the paper. This is a valid reason, and was probably once unimportant, but in climate science it is no longer the case that earth science is only of arcane interest; it is no longer a matter of 'gentlemen scientists' fossicking about with rocks and fossils (like some character in a Victorian novel), but an area of highly-politicized science. Moreover, any communication between authors and referees is only likely to further the process of noble cause corruption as *ad hominem* factors are more likely to take wing when the identity of the author is known to the reviewer. And, without the identity of the referee also being made known, they can be granted either a free pass or a rough ride with relative impunity, subject only to the ethical commitment of the editor.

Editors thus have considerable discretionary power, both in the selection of referees and in acting upon their reports. We have already noted how in the cases of the Huang *et al.*, paper and in the avoidance of the established authorities on satellite records in Spencer and Christy that *Nature* (in both cases) and *Nature* and *Science* (in the latter case), two leading general science journals appear to be favouring particular scientific conclusions. This appearance is not helped by the fact that both of them have shown a tendency to publish 'helpful' scientific papers on the eve of each annual meeting of the Conference of the Parties of the Framework Convention on Climate Change. McIntyre and McKitrick also suffered a similar fate to Huang *et al.* at *Nature.*

In addition to *Nature*, the leading US journal *Science* has also become political. For example, it published a paper in December 2004 on the eve of the 10th Conference of the Parties to the FCCC which purported to show that there was a unanimous scientific consensus on the anthropogenic causes of recent global warming (Oreskes, 2004). The author, Naomi Oreskes, claimed to have examined the abstracts of all 928 papers generated by a computer search of the ISI Web of Science produced between 1993 and 2003 using the keywords 'climate change' and found unanimity. The paper was palpable nonsense, as could quickly be verified by a replication of the search – a test any referee or editor could have subjected the paper to, had they bothered, and had they been at all sceptical as to the claim. Any one with the slightest familiarity with the field should have guessed that fewer than 100 papers per year was a laughably low figure, yet *Science* published the paper, and at a politically convenient time. In the UK, it was cited by the Royal Society and the government's Chief Scientist Sir David King as justification for immediate action. But then King, having endorsed the science in the disaster movie *The Day After Tomorrow* and having warned us that we might all have to live in Antarctica in future, was not the calmest voice of reason.

But a search of the ISI database using 'climate change' produced 12 000 papers, and Oreskes was forced to admit after science journalist David Appell (the owner of the blog where Mann had first mounted his defence) challenged her on his website (within 12 days of publication)[23] that she had used the three keywords 'global climate change', which reduced the return by an order of magnitude. *Science* published a correction by Oreskes (Oreskes, 2005), but it refused to publish a letter from Dr Benny Peiser which showed that her numbers could not be replicated, and another from Dr Dennis Bray reporting a survey of climate scientists showing that fewer than one in ten considered that climate change was *principally* caused by human activity.[24] Dr Bray told the UK paper the *Sunday Telegraph* that *Science* had informed him his paper 'didn't fit with what they were intending to publish'. In the case of Dr Peiser, *Science* used the familiar technique of asking him to shorten his letter, which he did, only to then have it rejected on the curious grounds that the 'basic points' of the letter had already been widely dispersed over the Internet – a point Peiser disputed and sought evidence for. *Science* thus avoided the embarrassment of printing a letter which drew attention to its lax review processes for politically convenient papers.

But the Oreskes paper was still being cited in May 2005 by Sir David Wallace, vice-president of the Royal Society in a letter asking the media 'to be vigilant against attempts to present a distorted view of the scientific evidence about climate change . . .' He added: 'I hope that we can count on your support.' And: 'There are some individuals on the fringes, sometimes with financial support from the oil industry, who have been attempting to cast doubt on the scientific consensus on climate change' (Collins, 2005).

Such is the naïveté of much of the political critique by the environment movement that they mistakenly attribute the opposition of ExxonMobil to Kyoto to a general opposition to any action on climate change. Exxon did oppose Kyoto, but largely because it did not like the policy approach and especially its implications for US–EU competitiveness. It is sceptical about climate change science, while at once spending hundreds of millions of dollars responding to the issue as a corporation. But, as the largest private gas company globally (and the second largest gas supplier after the Russian state-owned monopoly Gazprom), Exxon stands to be *advantaged*, because the easiest way to decarbonize economies rapidly is to switch from coal-fired electricity generation to combined-cycle gas turbine generation, which reduces GHG emissions by around 60 per cent. (It is unlikely Exxon's petroleum interests will be substantially harmed by decarbonization policies over commercial time-frames of 20 or so years, given global demand for transport fuels and the supply outlook.) Gelbspan (1998, 2005) has documented the funding Exxon has provided to climate change sceptics and to think tanks and institutes with which they have some affiliation ($80 000 to the Fraser Institute, where McKitrick is a fellow, but not for climate change). Those looking for cases of venal corruption usually overlook the possibility that sceptics might hold their views sincerely, and (not unnaturally) economic interests might support them because of such views.

Gas producers and the nuclear industry receive a 'free pass' in climate change politics, because of a 'Baptist and bootlegger coalition' in which economic interests are empowered by the moral cloak of a noble cause (Yandle, 1989, 2001). There are examples of such coalitions (in addition to gas and nuclear interests) supporting decarbonization policies (for example, to cite but one headline, 'Wind farm expansion as manufacturer is shown to be Labour donor' *Daily Telegraph*, 23 May 2005). But what is often overlooked in this is that the noble cause, as Muller notes, leads us to grant a free pass to science that is less than sound, which makes the adherence to high standards in the conduct of science in these areas all the more important.

A particularly egregious study of news coverage of climate science, arguing that by attempting to provide balance news media end up being biased (by including minority views), demonstrates strangely partial scholarship (Antilla, 2005). Not only does it cite the flawed Oreskes (2004a) paper as its authority for the statement that the view that there is substantive disagreement in the international scientific community is false, but it then seeks to discredit reputable dissident scientists such as John Christy (the leading authority of satellite records) by pointing out that he is listed as being on the global warming panel of the Independent Institute to which ExxonMobil donated $10 000 in both 2002 and 2003. This game of 'six degrees of

ExxonMobil' elevates the genetic fallacy to research design, fallaciously suggesting that Christy's scientific views reflect this support (rather than the possibility of the reverse causation). If we were to accept this as an argument, we would have to dismiss the views of Stephen Schneider and Paul Ehrlich on the grounds that Exxon also donates the rather more substantial sum of $10 *million* annually to Stanford University (where they both work) for climate change research. Even worse, this author refers to a report by this think tank 'based on the research of Christy and others, which appears to be non-peer-reviewed.' This slur ignores the fact that Christy's work has been published in the best peer-reviewed journals, and comes on the same page as a reference to a story in the non-peer-reviewed magazine *The Ecologist* – which is presumably acceptable because its politics are sound.

In addition to the politics that surround climate change science, the prominence of computer models in climate science is surely also a factor in producing uncertainty. The models themselves are often judged by how well they agree with other models – a rather weak test, since they are likely to share many lines of code. But model results are frequently accorded a 'concreteness' that is more than a little worrying. We can illustrate this by reference to a significant paper – often cited to downplay the effect of solar variability in climate variability – which demonstrates the circular and virtual nature of much climate science. Crowley (2000) concluded that solar factors are minor compared with GHG forcing, but his study used *model-generated* estimates of 'radiative forcing' due to solar flux, atmospheric aerosols and CO_2 levels, to see how well they could explain the Hockey Stick curve produced by Michael Mann – itself a reconstruction of past temperatures using various proxies for temperature.

Crowley used an Energy Balance Model (EBM) to explain temperature changes, but the model must embody assumptions about how CO_2, solar flux, etc., influence temperature, which Crowley explained thus:

> The radiative damping term can be adjusted to embrace the standard range of IPCC sensitivities for a doubling of CO_2. The EBM is similar to that used in many IPCC assessments and has been validated against both the Wigley-Raper EBM and two different coupled ocean–atmosphere general circulation model (GCM) simulations. All forcings for the model run were set to an equilibrium sensitivity of 2°C for a doubling of CO_2. (2000, p. 272)

Models are being used to validate models, which in turn are being used to reproduce proxy reconstructions. The 'data' used by Crowley (and others in this 'signal detection' genre) is generated by a model that includes an assumption that CO_2 is a powerful greenhouse gas (doubling is taken to increase mean global temperatures by 2°C). This model-generated 'data' is then effectively used to test whether the climate is sensitive to CO_2.

As we have noted in relation to the paper by Huang *et al.*, there is a discrepancy between the records computed from boreholes and those compiled using proxies, the most notable of which is that of Mann *et al.* Mann and Schmidt (2003) claimed to provide a partial explanation for the discrepancy between temperatures as computed from borehole measurements and proxy reconstructions such as theirs, using a model simulation (GISS ModelE GCM) of the latter half of the twentieth-century. Chapman *et al.* (2004) attempted to account for the same difference between borehole and proxy records, but using actual measurements of GST and SAT at observatories and obtained the opposite results, and argued that the differences were based in a bias resulting from 'selective and inappropriate presentation of model results by Mann and Schmidt'. Understandably, Schmidt and Mann (2004) rejected the criticism of their paper, but were unrepentant in their use of models. Mann has acknowledged other errors. After the publication of a reconstruction by the Hockey Team of temperatures using borehole data (Mann *et al.*, 2003c), an error was pointed out (Pollack and Smerdon, 2004), and a correction was issued – with Mann forgoing lead authorship in favour of his junior colleague Rutherford, who had earlier been been assigned responsibility for apparently sending McIntyre the wrong data (Rutherford and Mann, 2004).

It is possible to find problems with most of the IPCC depictions of future climate – or present climate, for that matter. There continue to be problems with the reliability of the historical temperature record, which has been constructed by adjusting data to remove unreliable sites and adjust for the urban heat island effect resulting from both energy use in cities and the greater absorptive properties of cities compared with rural or natural locations. Pielke and Matsui (2005), for example, published a paper suggesting a possible warming bias in urban temperature records, again with a paper that had been rejected by *Nature* detailing criticisms of a paper in *Nature* purporting to set aside concerns over the urban heat island (Parker, 2004). So the temperature record during the period for which there is an instrumental record depends upon adjustments to the raw data and is open to question, just as is the climate history of the past millennium accepted by the IPCC. And there are alternative explanations of recent warming, such as the perhaps obvious one of increased solar activity, with research indicating a level of solar activity as measured by sunspots since 1994 unparalleled in the past 100 years (Usoskin *et al.*, 2003). Such challenges are, however, explained away by the climatology establishment.

Many other key data are similarly subject to adjustment. The high values of CO_2 concentrations from ice cores are manipulated, with high values discarded as being contaminated, and cross-validated against a selective reading of historical analyses. There might be good reasons for adjusting data in this way, but any adjustment allows opportunities for subjective factors to intrude into the analysis; even relatively innocent adjustments can be made in the light of preferred theories, rather than necessarily to support the political agendas of researchers, but both are possible, and call attention to the need for transparency in research.

There are numerous examples of modelling results being used to 'verify' climate science theories. The 21 October 2005 issue of *Science* carried a paper which drew attention to the risks to sea level rise posed by what the authors saw was a decline in the mass of the Greenland ice sheet. The authors then went on to state that the estimates of ice loss were 'broadly similar' to the 'results' from a model used to simulate the surface mass balance of the Greenland ice sheet from 1991 to 2000. The estimate from the surface mass model was 78 Gt/year while their estimate from repeat altimetry and atmospheric and runoff *modelling* was 54±14 Gt/year (Alley *et al.*, 2005). The surface mass model estimate was therefore outside the error range of their estimate, and about 45 per cent higher, which represents a strange notion of similarity. Ironically, one day before this paper appeared in print, another was posted on *Sciencexpress* stating that previous mass balance work on the ice sheet was based on altimetry sampling that was uneven in space and time, such that the previous surface–elevation data sets that had been analysed had been discontinuous and relatively short. Johannessen *et al.* (2005), in this new paper, presented a much more comprehensive data set showing that, while there had been a small decline at low elevations, there was substantial positive ice growth above 1500 metres which more than offset the low elevation decline, making for an average annual *increase* in ice mass over 1992–2003 of 5.4±0.2 cm. (There has been a cooling trend at the summit of the Greenland ice sheet (Chylek *et al.*, 2004)).

In a similar vein, a paper published in *Science* in 2006 found evidence (using satellite measurements of gravity) of a decline in ice in Antarctica (Velicogna and Wahr, 2006), but it was based on only 34 months of data and had as its starting point a time of exceptionally thick ice as recorded by other means in the period 1982–2003 (Davis *et al.*, 2005). By selecting only 34 months of data, the paper ignored the evidence that ice mass had been growing for 20 years. Media coverage was predictable, with the *LA Times*, *New York Times* and *Washington Post* all running coverage, on 3 March, largely uncritical of the study's limitations, with the *Post* headline saying it all: 'Antarctic ice sheet is melting rapidly: New study warns of rising sea levels'. (The West Antarctic ice sheet has been declining, but the much larger East Antarctic ice sheet appears to have been gaining mass.)

Much of the science of climatology involves unsatisfactorily short runs of data which is subject to manipulation. For example, Kerry Emmanuel published a paper in *Nature* in July 2005 in which he showed that there had been an intensification of hurricane intensity over the preceding 30 years. These findings were inevitably linked to climate change and took on special meaning when Hurricane Katrina struck New Orleans shortly after they were published. But data on hurricanes have only been gathered for 150 years, and only 60 years of wind speed data exists (from the commencement of flying into hurricanes). Previously, wind speeds were extrapolated from damage and from data collected on ships and on land, and some remote storms might have not been recorded at all. The use of satellites since the 1960s has improved data and meteorologists produced formulae for calculating wind speed in order to smooth out the historical record.

Emmanuel used these formulae to adjust for years when he considered wind-speed data were unreliable, particularly in Atlantic storms from 1949–1969, when it is considered that speed was overestimated, and then cubed wind-speed to reflect the exponential increase in damage with wind-speed. This process, of course, also accentuates any changes in data. As is immediately obvious, this analysis depends upon the considerable manipulation of data rather than the use of observational data, and we have far too short a run to make any statement about trends. As a consequence there is considerable controversy over these findings, with bitter divisions apparent between Emmanuel and those such as (on the one hand) Greg Holland, a colleague at NCAR of Kevin Trenberth, whose statements linking hurricanes to global warming led to Chris Landsea resigning as a lead author of the IPCC, and (on the other) Dr William Gray, perhaps the most prominent US hurricane expert (Bauerlein, 2006). It is of interest to note that a consensus document submitted to the WMO Commission for Atmospheric Sciences in February 2006 (by a panel which included Emanuel, Holland and Landsea) stated that there was no evidence of an increase in cyclone frequency and that scientists were 'deeply divided' over any trends in intensity (WMO, 2006). This controversy continues, with Landsea *et al.* (2006) attributing Emmanuel's finding to improvements in data sources, whereby some older intense cyclones were simply not counted because of inferior technology (only two geostationary satellites before 1975, compared with eight now, for example). Regardless of the outcome of this controversy, it has been egregious in the extreme to attribute Hurricane Katrina to climate change, since this was a case of only a Category 3 hurricane impacting an extremely vulnerable city.

Such factors drive climate scientists to a reliance upon models. As we noted at the beginning of the chapter, Stephen Schneider has suggested that the use of historical situations is 'essential to test the tools used to make future estimates', and he calls this a 'systems version of "falsification"'. But proper falsification is possible – if we are prepared to wait – because the model results can eventually be tested against future changes in climate: it is only our pre-existing belief in the need for concern over the urgency of policy action that prevents us from awaiting the results of such falsification. There was apparently once a saying used at IBM that the problem with simulation was that it was like masturbation: if one indulged in it too often, one began to mistake it for the real thing. It is only our belief that the 'real thing' in climate terms is likely to happen and will be disastrous that leads to a preference for a 'systems version' of falsification, rather than the real thing.

The political context within which climate science is conducted – that there is an urgent problem which precludes waiting for observational falsification of models – corrupts the conduct of climate science by privileging modelling as an activity. As Hansen (2002, p. 346) has noted, this is actively encouraged by the operation of the IPCC: 'One problem with the IPCC reports is that each report produces new (and more numerous) greenhouse gas scenarios with little attempt to discuss what went wrong with the previous ones.'

Reliance upon models, the workings of which and the assumptions inherent in, are known to only a few, facilitates the intrusion of subjective factors into research. The absence of a well-developed tradition of data archiving, transparency and disclosure of methods, together with the existence of tight networks of researchers in close association thanks to modern communications, also mean that the usual checks and balances that operate in other areas of science are lacking and can provide no assurance of quality. Favourable findings are likely to be granted a free pass because they support the political cause, while inconvenient findings are likely to be met by attempts to explain them away.

An example of this came with the publication of a paper in *Geophysical Research Letters* in August 2006 which reported some surprising results from the deployment of a global array of 2500 ARGO floats, designed to descend to various depths to measure ocean temperature and salinity, and then surface periodically and transmit their data via satellite. The ARGO system had been hailed as likely to provide the best data documenting changes in ocean temperatures, expected to be a warming – crucial, because the oceans hold most of the energy in the climate system, acting as a kind of 'flywheel'. The problem was, the data indicated a cooling, with the oceans losing between 2003 and 2005 about one-fifth of the energy they had accumulated over the previous 50 years. Sydney Levitus, who headed a research team that had found warming between 1955 and 1998, understandably, stated: 'What it does tell us is that we still don't sufficiently understand how the global climate system works' (Schiermeier, 2006, p. 854). Climate modeller (and Mann co-author and fellow *RealClimate* contributor) Gavin Schmidt also indicated the finding showed the coupled ocean–climate models did not capture 'intermittent fluctuations', while adding that a short-term cooling 'blip' couldn't say much about the climate system in general. The heat could be hidden at greater depths in the ocean, or more likely (said Schmidt) it might have escaped into the atmosphere and out into space – though it was noted by critics that a corresponding change in the earth's radiation budget had not been observed (Schiermeier, 2006). This typified the behaviour scientists exhibit when confronted with contrary facts, seeking to explain them away (Barber, 1961), though the greatest disconnect was with models, not with observations. As it happened, a subsequent correction reported data problems with the ARGO system, with neither warming, nor cooling evident.

The defence of the new climate science orthodoxy can be vigorous, to say the least. Jan Veizer and Nir Shaviv published the results of research which suggested an alternative explanation of climate variability – that a complex interplay of variable cosmic rays and solar energy affected the water cycle of clouds, rainfall, surface evaporation and transpiration by plants, and only then did carbon dioxide become involved, amplifying changes in temperature set off by those primary forcing agents. Veizer and Shaviv published in *GSA Today*, the journal of the Geological Society of America. An attack on this piece was launched, not in *GSA Today*, but in *EOS*, the journal of the American Geophysical Union, by a group of 11 scientists, headed by Stefan Rahmstorf (a fellow contributor at the *RealClimate* website with Michael Mann). Rahmstorf claimed that they had not submitted their response to the paper to *GSA Today* (as scientific courtesy would require) because he had been told the journal had a policy against publishing such articles. He clearly did not bother to check its editorial policies, because it had no such policy and would have welcomed such a submission *but* would have offered Veizer and Shaviv a right of reply (Calamai, 2004).

Remarkably, the first direct quotation in the Rahmstorf critique was not from the scientific paper by Veizer and Shaviv, but from the press release issued by the Hebrew University of Jerusalem, where Shaviv was now working. Rahmstorf was with the government-funded Potsdam Institute for Climate Impact Research, a modelling centre, and he explained his group of critics acted because of the publicity the Veizer and Shaviv paper had received and because it 'had so much attention with the wrong audiences', as one of Rahmstorf group, Andrew Weaver put it. Rahmstorf had even pressured Ruhr University (where Veizer held a joint appointment with University of Ottawa) into removing from its website a graph associated with the press release about the research. Rahmstorf stated: 'I perceive this graph as a way of misleading the public. It's used by various lobby groups that are opposed to reducing carbon dioxide emissions' (Calamai, 2004).

Again, the overwhelming concern was with the uses to which the science might be put, not its rigour.

As we have seen, the models and the scenarios driving them are extremely limited, and are replete with examples of poor scientific practice. The reliance upon models and emission scenarios lends a veneer of credibility to warnings of looming disaster, but it is highly questionable whether they provide a better understanding of the problem of climate change than science based solely upon observational science. In 1990, three distinguished scientists, Robert Jastrow, William Nierenberg and Frederick Seitz (1990), provided an alternative to the IPCC computer generated outlook of a warming of 1.5 to 4.5°C by 2100. They assumed the range of the temperature increase in the twentieth century was 0.3 to 0.6°C, and assumed that all this was due to a roughly 50 per cent increase in GHGs from pre-industrial levels. If we then assumed a further 50 per cent increase over pre-industrial levels over the next century we could see an increase of 0.6 to 1.2°C, or a range of 0.8 to 1.2°C assuming a correction for a lag in warming not yet observed due to heat in the oceans. They then added a margin of 0.4° either way to allow for natural climate variability, to give range of 0.4 to 1.8°C.

This rather simple set of assumptions and calculations turns out to have provided a remarkably accurate prediction of what happened in the years after, because the trend in surface temperatures in the late twentieth century has been running at about 0.17°C per decade. If all of this were to be ascribed to anthropogenic causes, we could expect another 1.7°C over the next century. This is the ballpark figure of the amount of warming predicted by James Hansen, whose testimony to the US Congress in the midst of a hot, dry summer in 1988 is attributed with providing the political impetus for the issue. As we have noted, Hansen has been critical of the IPCC for its 'predilection' for exaggerating growth rates of population, energy intensity and pollution, and for its 'failure to emphasize data', allowing scenarios to be used with rates of increase of GHG emissions double those observed during the 1990s (Hansen, 2002, p. 437).

Hansen is a prominent pro-warming scientist, and a fierce critic of the Bush administration – even speaking against the administration that employs him in the key state of Iowa during the 2004 election campaign – but is also a strong critic of the virtual world of climate science encouraged by the IPCC, with its preference for models over observations. But Hansen's impeccable credentials as a climate scientist did not exempt him from a reaction usually reserved for dissident voices when (with others) he published his 'alternative scenario' as a policy response to the problem (Hansen *et al.*, 2000). The Hansen Alternative Scenario still calls for limitation of carbon dioxide emissions, but de-emphasizes carbon dioxide, relatively speaking, by drawing attention to the fact that climate forcing also results from other factors which can be mitigated more cheaply, or with substantial co-benefits, and with greater technical ease, than carbon dioxide. It suggests that the Kyoto Protocol inappropriately focuses on carbon dioxide, to the neglect of other climate forcing agents.

While Hansen considered *Science* provided a factual summary of their paper, *Nature* made several 'misstatements' and quoted only critics of the paper, in what was published as a 'News' article (Smaglik, 2000), whereas Hansen considered it to be so opinionated that it should have been published as an editorial. He later stated in an open letter on the matter (written 'When [he] had difficulty publishing a response in *Nature*') that 'This was made clearer to [him] when [he] submitted a "letter to the editor" to correct their misinterpretations, because they objected to [his] letter and edited it in a way that altered the meaning.' Hansen was also criticized by the Union of Concerned Scientists (UCS), which sent its members an 'Information Update' discussing the paper, arguing that it was 'controversial, potentially harmful to the Kyoto Protocol, and not a helpful contribution to the climate change discussion as it "may fuel confusion about global warming among the public"' (Hansen, 2000). He later told a Senate Committee that the paper received a similar response from the IPCC, which did not include it in the TAR (partly because it appeared so late). Asked 'Do you feel that your results were reviewed and properly considered as part of the IPCC process?', Hansen replied with a straight 'No,' and then added: 'The only IPCC "review" of our paper was by the IPCC leaders . . . who saw our paper as potentially harmful to Kyoto discussions' (Hansen, 2001).

POLITICAL SCIENCE

The attack on McIntyre and McKitrick was not therefore an isolated case, though it was certainly conducted more vigorously, perhaps because they were not mainstream scientists. We can point to similar treatment meted out by scientists subscribing to the dominant consensus (and journals like *Nature*) to more mainstream scientists such as Veizer and Nir Shaviv, Spencer and Christy, and even to Hansen when he proposed something contrary to the IPCC consensus in favour of the Kyoto Protocol. The IPCC similarly impugned the character of Castles and Henderson in an effort to repel the criticism they mounted against the SRES – which even Hansen considers faulty because it contains scenarios for emissions for the 1990s in excess of those actually recorded. Similarly, the threat posed to climate modelling by the ARGO observations showing *cooling* was dealt with by a pre-emptive news story which sought to explain away these inconvenient data: doubts were raised because they did not agree with models or with sea level rise measured remotely, and there were even suggestions that the heat was still there, but hidden at depths, or had escaped to space through the atmosphere (seemingly without being detected).

The controversy over the Hockey Stick excited political interest in the USA, with the Chairman of the House Energy and Commerce Committee, Representative Joe Barton, writing to the Hockey Team, posing a list of questions and asking for Mann to disclose his source code, which he eventually did. For this, Barton was accused of intimidating the Hockey Team by National Academy of Science (NAS) President Ralph Cicerone, himself a climate scientist. Barton instigated an inquiry by his committee, and commissioned an analysis of the Hockey Stick by a panel convened by Professor Edward Wegman, a distinguished statistician from George Mason University (Regalado, 2006). Barton's inquiry was criticized by several scientific societies as well as by his fellow Republican, Representative Sherwood Boehlert, Chair of the House Committee on Science, who proposed to Barton a joint approach to the NAS to ask for a review. Barton failed to respond, perhaps smarting from Boehlert accusing him of making a blatant effort to intimidate global-warming researchers. Boehlert then made an independent request to the NAS in November, which a spokesman for Barton's committee stated was not likely to address all of Barton's concerns (Regalado, 2006). As it happened, this proved correct, as the NAS panel failed to address the important question of whether researchers should disclose fully their data and methods.

Once the results of the Wegman and NAS panels were in, it was apparent that the NAS panel was as much an exercise in damage limitation as a genuine attempt to enquire into the episode, because (while it upheld practically every criticism McIntyre and McKitrick had raised), the NAS used remarkable 'spin' to package the findings in the best way possible to minimize the harm to climate science and the cause of climate change politics. It stated in its press release that the panel had confirmed that the present climate was the warmest in the past 400 years, and many media sources picked this up and gave it prominence as news. In fact, the report restored the *status quo ante* and Little Ice Age, because (rather than being news) that was the overwhelming consensus position *before* the publication of the Hockey Stick. Of the climate before 1600, it simply said the Hockey Stick was 'plausible,' pointing to numerous 'independent' studies that confirmed it – neglecting the fact (as McIntyre and McKitrick were quick to point out) that most of these repeated the same mistakes of Mann *et al.*, and neglecting the obvious point that the authors of these studies, rather than being independent, were close research associates of Mann and his team (as Wegman was to show with a network analysis in his report).

Remarkably, the NAS panel failed to probe the question of Mann's 'CENSORED' file, which indicated he had performed the analysis without the crucial bristlecone pines, knew the result was dependent on their inclusion, and neglected to mention this. And the panel also declined to address the crucial issues of disclosure and transparency.

While McIntyre was later to state he was largely satisfied with the NAS Panel findings (while regarding the report as somewhat 'schizophrenic'), and von Storch was to remark it was about the best that could have been hoped for, given the reputations and politics riding on it, it seems as if the panel had been composed so as to minimize damage, from its restricted scope and selection of members. McIntyre and McKitrick had exercised their right to make a submission on the composition of the NAS panel. They specifically objected to the appointment of three panel members on the grounds that (while suitably qualified) there were conflicts of interest or evidence of pre-existing views that might bias their deliberations: Dr Otto-Bliesner, Dr Nychka and Dr Cuffey. Dr Otto-Bliesner was a frequent co-author with Dr Caspar Ammann and was his direct supervisor at University Corporation for Atmospheric Research at the University of Colorado (UCAR). Ammann had been the source of a 'media advisory' in May 2005 defending Mann and the Hockey Stick that had been relied upon in material presented to the US Congress by Sir John Houghton of the IPCC, by Mann, and by the European Geophysical Union. The paper underlying this press release had been accepted for publication in *Climatic Change* by Schneider without reference back to McIntyre, who as a referee had recommended against publication. As of May 2007 it was yet to appear in print yet it was included in the Fourth Assessment Report of the IPCC in seeming breech of their rules against including unpublished material. While touted as independent verification of the MBH Hockey Stick, Ammann was a *RealClimate* contributor and had been a doctoral student under both Mann and Bradley. Dr Otto-Bliesner had also co-authored a paper with one member of the Hockey Team (Bradley) and served on a committee with him. Dr Nychka was also a UCAR employee and was currently collaborating with both Ammann and Mann, while McIntyre and McKitrick objected that Dr Cuffey had recently made statements indicating a possible bias in an op-ed piece in the *San Francisco Chronicle.*

UCAR was well represented on the NAS Panel, with three others on the panel (Chairman North, Turekian and Dickinson) either present or past UCAR trustees. The NAS made a late addition of a statistician to the panel: Dr Peter Bloomfield, of North Carolina State, who had co-authored several publications with Nychka and had acted as a consultant to the Hockey Team on one paper. While the panel included John Christy, therefore, its composition goes some way towards explaining what McIntyre called its 'schizophrenia'.

What was interesting about the NAS and Wegman panel reviews was that, while Wegman performed analyses to check the validity of the McIntyre and McKitrick claims (which it upheld), the NAS panel performed no additional research and chair North subsequently told a seminar at his home university (Texas A&M) that they just 'winged it,' stating that that was 'what these panels did'. The North Panel concluded that twentieth century warming was supported by borehole temperature measurements, the retreat of glaciers and other observational evidence 'and can be simulated with climate models' (though what this last statement added to observational science is not clear). It also found that large-scale surface temperature reconstructions yielded 'a generally consistent picture of temperature trends during the preceding millennium' including a Medieval Warm Period centred around AD 1000 and a Little Ice Age centred around 1700. The smooth handle of the Hockey Stick was thus overturned and much of the previous variability restored.

The panel did state that the basic conclusion of the MBH paper that late-twentieth-century warmth was unprecedented had been supported subsequently by a large array of other studies (most of which, as we noted, McIntyre was quick to point out, included the flaws such as bristlecone pines and computational steps the panel had ruled against), but not all proxy records indicated that recent warmth was unprecedented. The panel report included a graph featuring many of these studies, from which they had 'winged it'. One of these was of some interest: a study of borehole temperatures since 1500 by Huang *et al.* (2000), which showed a rise in temperature. But given that the panel's report tended to support the MBH reconstruction since 1600 (or, rather, gave the impression that it did), it is interesting that Huang *et al.*, had explicitly compared their results against three proxy reconstructions including MBH and Mann co-author Jones (Overpeck *et al.*, 1997; Jones *et al.*, 1998; Mann *et al.*, 1998). The borehole record (considered by many to be the most accurate method because it is a direct measure rather than a proxy) found MBH to be the most divergent from boreholes, understating the depth of the LIA by about 0.5°C, followed by Jones *et al.*, with Overpeck *et al.*, the closest to the borehole record.

None of this was apparent from the NAS panel report, because the Hockey Stick was not included in the 'spaghetti graph' they presented and the extent of its error was not made obvious. Significantly, also missing was the earlier graph by Huang *et al.* (1997) that had shown not just a clear Little Ice Age, but also a definite Medieval Warm Period. This (and the damage limiting press release) allowed Mann and his supporters to claim victory. *Scientific American* (28 June 2006) headlined its coverage 'Academy affirms hockey-stick graph; But it criticizes the way the controversial climate result was used.' Conveniently overlooked was the fact that this misuse had been committed primarily by Mann himself, as lead author of the IPCC chapter that reported his own science.

CONCLUSION

Hans von Storch is one scientist who has occupied a difficult middle territory in the controversies of climate science we have examined here: resigning as Editor-in-Chief of *Climate Research* on a point of principle over the publication of a paper critical of Mann *et al.*; himself an author of research critical of Mann *et al.*; yet one who accepts that the world is experiencing anthropogenic climate change which requires a policy response. He has praised the work of McIntyre and McKitrick and the contribution they have made to climate science. In testimony to Rep. Barton's Congressional Committee in July 2006 he lamented what he saw as the gate-keeper role played by a small number of scientists in the field of historical global climate reconstructions, 'where a small group of scientists has exerted an undue control of the entire field' (von Storch, 2006, p. 8).

While he saw the close nature of the Hockey Team as working to restrict publication of dissident views, von Storch was critical of the lack of disclosure which prevented reproducibility and was critical of the standards of both *Nature* and *Science*, which he saw as facilitating lack of disclosure through practices such as the short length of article they required (which limited the amount of detail provided) and their focus on stories with strong media appeal (since they were commercial operations that relied upon sales). He also considers that most of environmental science, including climate science is, an example of 'post-normal' science (Bray and von Storch, 1999): loaded with a high degree of uncertainty on an issue of great practical importance. With such science, the boundaries between science and value-driven agendas get blurred, representatives of NGOs are thought to know better about the functioning and dynamics of systems than scientists, parliamentary committees delve into the technicalities of science, amateurs engage in the technical debate, and some scientists try to force solutions on policy-makers and the public.

The argument here is that it is the virtual nature of post-normal science that is especially dangerous. The extensive reliance upon models and the significant manipulation of their source data creates the danger of virtuous corruption, just as the values of those who wish to push policy prescriptions on to policy-makers and the public can (even if inadvertently) contaminate the conduct of their analysis. Research protocols in other areas of science (such as medicine) anticipate such a possibility and work to minimize it. Climate science would seem to have an underdeveloped sense of the importance of quality assurance safeguards such as double-blind experiments and double-blind reviewing of research papers.

In 2005 Margaret Beckett, UK Secretary of State for the Environment, Food and Rural Affairs, announced funding of £12 million to change public attitudes to climate change. She stated that 'We need people to understand that climate change is happening. . . . It is essential to deepen popular understanding and support for action on climate change' (DEFRA, 2005). Leaving aside the point that there is something quite bizarre (and slightly undemocratic) about a government spending its taxpayers' money to change their minds, such efforts could well be undermined by the extensive reliance upon models in climate science. None of them have yet predicted the failure of the oceans to warm, and none of them predicts a decline in mean global temperatures, yet the chaotic nature of the climate system suggests that such are entirely possible, and entirely consistent with the possibility of climate change.

This raises the question of whether the risks of corruption of science, which further contribute to a loss of credibility, are worth the benefits of the reliance on modelling, since Hansen's view of the likely warming is largely consistent with that derived by Jastrow, Nierenberg and Seitz from observational premises in 1990. Reliance on modelling gives rise to problems: assumptions must be made at every stage of model construction; their complexity limits the possibility that they can be audited; they must omit factors in order to simplify reality and these omissions can prove crucial in unforeseen ways; even simple measurements can be affected by the desires and expectations of researchers; models are often tested against essentially the same observations they were built from; they do not account for chaotic changes, such at the 1976 Pacific Climate Shift (see Namias, 1978). If models are to be used, it is essential that this occurs within a climate of open scepticism and contestation. Lahsen (2005) found that climate modellers are often least able to see the problems with their models.

Such questions tend to be lost in the heated politics which surrounds the science of environmental problems, and there have been few examples of more heated science than that which greeted the publication of a book by Bjorn Lomborg which dared to question the bleak nature of the prevailing environmental outlook. This is the subject of the next chapter, where we shall see that questions such as the impact of climate change, with which (*inter alia*) Lomborg dealt, both exposed a world of species–area models, fed by climate models, fed in turn by emissions scenarios. The result was the activation of all the political defence mechanisms of environmental science.

NOTES

1. www.realclimate.com. MBH also responded with specific counters to the criticisms of McIntyre and McKitrick; see: Mann *et al.* (2003b).
2. These details are drawn from 'M&M Project Update' September 2004, http://www.uoguelph.ca/~rmckitri/research/fallupdate04/update.fall04.html, accessed 10 October 2004.
3. Details were recorded by McIntyre on Climate Audit on 28 March 2006, www.climateaudit.org/?p=607#more-607, accessed 7 April 2006.
4. Briffa and Osborn, 2002; Briffa *et al.*, 2001; Briffa *et al.*, 2002; Briffa *et al.*, 2004; Osborn and Briffa, 2002.
5. Cook *et al.*, 1995.
6. Briffa *et al.*, 2001; Briffa *et al.*, 2002; Jones *et al.*, 1988; Jones *et al.*, 2003.
7. Cook *et al.*, 1995.
8. Cook *et al.*, 2004; D'Arrigo *et al.*, 2005.
9. Cook *et al.*, 2004.
10. Mann *et al.*, 2003a.
11. Cook *et al.*, 2004; D'Arrigo *et al.*, 2005.
12. Briffa *et al.*, 2001; Jones *et al.*, 1998.
13. Jones and Mann, 2004.
14. Jones and Moberg, 2003.
15. Jones *et al.*, 2003.
16. Cook *et al.*, 2002.
17. Jones and Mann, 2004.
18. Jones and Moberg, 2003.
19. McIntyre (2005).
20. Climate Audit, 15 October 2005.
21. Climate Audit, 18 July 2005.
22. www.sciencepolicy.colorado.edu/prometheus, accessed 8 February 2005.
23. http://davidappell.com/archives/00000497.htm, Accessed 23 February 2005.
24. *Sunday Telegraph*, 1 May 2005.

4. Defending the Litany: the attack on *The Skeptical Environmentalist*

> [In] recent years it is apparent that the enormous significance attached to climate in earlier times was somewhat overdone because people failed to see that many of the effects that seemed to be due to climate in itself were really due to quite different factors only indirectly related to climate. . . . This is not to deny that climates have their own specific qualities or that nobody, if they could help it, would choose to live in Ghana, that dreadful Turkish bath supplied by nature, or in Death Valley or even Alaska, but what has this to do with health? The answer is: Nothing at all (*see* Hypothermia).
>
> *Pears Medical Encyclopaedia* **J.A.C. Brown**, compiler, (1969), pp. 118–20

Climate change is a policy problem which inherently involves making decisions under uncertainty (Kellow, 2005). As we saw in the previous chapter, there is much uncertainty in our understanding of past and present climate, let alone how the climate might change over the next century or so. Our case studies of the Hockey Stick controversy and the SRES scenarios illustrate that such science is conducted within a context of vigorous politics, where the findings matter, and the nobility of the cause of 'saving the planet' from catastrophic climate change results in a preparedness to accept lower standards of proof, and even 'science' which it is hoped will lead humanity to 'do the right thing'.

There are some claims made in the popular media which, frankly, will look more ridiculous with the passage of time, and will likely cause not a little embarrassment for some of the people making them. Antarctica is likely to be the only habitable continent by the end of the century, according to the UK government's Chief Scientist, Sir David King, on the eve of an important political meeting (*Independent*, 2 May 2004). The future of Christmas was under threat in one press release, because global warming would threaten reindeer habitat. Scotland is at risk of heatwaves, according to another (*Sunday Herald*, 15 January 2006). (One wonders whether Scots would see that as a problem or a blessing.) On the other hand, climate change could end the Gulf Stream and plunge Britain into icy cold.

All these prospects are made possible by the high level of dependence in climate science on models and on proxy data and reconstructions which allow unstated assumptions and values to intrude into analysis. David Demeritt (2001) has suggested that in the way in which climate science is socially constructed, politics gets built into science at the 'upstream end'. These dangers are well enough known in other areas of science, and are safeguarded against by a number of checks which appear largely to be absent from climate science: archiving and disclosure of data; declarations of conflicts of interest; double-blind methods in the conduct of experiments; double-blind refereeing; and so on. Such practices are not well-developed in climate science, but they are demanded in other areas of scientific endeavour where the stakes are high (including medical research and geological exploration). Yet the 'virtual' nature of much climate science – now used to justify public policy involving massive stakes – makes it all the more important that the reliability of science is maximized.

This problem is exacerbated because much of the science relating to the impacts of climate change compounds the problems of virtual science with more virtual science. Not only are the results of climate modelling driven by the IPCC SRES scenarios or a hypothetical doubling of carbon dioxide (a hypothesis initially used to assess the performance of the models, rather than because it was likely), but they in turn are then used to drive ecological models (such as the species–area model we examined in Chapter 2) to produce bleak pictures of biodiversity loss, species extinction or increased human deaths. Again, there is a preference for the results of such modelling over observations which present a less serious or even contrary picture, especially by political groups pressing for political action, but also sometimes by the scientists themselves.

Much of this modelling is of a questionable and decidedly reductionist character. Economic forecasts are used to produce emissions scenarios, which drive climate models, which drive models of impacts on human health or species–area models, which in many cases ignore the fact that the whole process began with substantial economic change – which provides the capacity for adaptation. The worst of these examples relate to deaths from 'tropical' diseases like malaria and dengue fever, because the distribution of insect vectors is supposed to be extended by climate change. These analyses ignore the fact that malaria is *not* a tropical disease, the anopheles mosquito having been endemic up to the Arctic Circle, and malaria endemic in Britain and much of Europe, only having been declared eradicated in the 1970s. The eradication of the disease came as the result of the very economic advances that, under the climate scenarios, will drive the emissions growth that will drive the climate change – yet the impact studies assume that the climate will change, but not the ability of the societies whose emissions will drive the change to introduce the socio-economic improvements, pesticide use, and so on that eradicated malaria from already-affluent countries.

One example of research of this kind projects various impacts into the future for: unmitigated emissions growth, stabilization at 750 ppm carbon

Table 4.1 Number of people at risk of falciparum malaria (millions)

	Unmitigated climate change	Stabilization @750 ppm	Stabilization @550 ppm
1990	4413		
2020s	6743	6890–6914	6923–6954
2050s	8072	8269–8330	8296–8292
2080s	8820	9076–9143	8994–9073

Source: Table IX, Arnell *et al.*, 2002, p. 439.

dioxide and 550 ppm carbon dioxide, and produces projections for populations 'at risk' to falciparum malaria that are larger than the current global population, and not far short of the latest UN projection for the level at which global population is expected to stabilize at about 10 billion at around 2050 (see Table 4.1).

Arnell *et al.*, acknowledge that 'much of the projected expansion of potential transmission areas is in developed countries, where public health infrastructure and other socioeconomic factors make it unlikely that actual transmission would occur' (Arnell *et al.*, 2002, pp. 437–9). But then one wonders why one might be bothered performing this exercise at all. Such modelling does, of course, allow others to ignore the qualifying socioeconomic factors and raise alarm that climate change could extend the range of disease vectors to cause actual deaths. One such example came with the publication of a consultant's report sponsored jointly by the Australian Medical Association (AMA) and the Australian Conservation Foundation (ACF) in 2005 which was reported under the headline 'No stopping deaths from climate change' (*ABC Science Online*, 22 September 2005). In this case the culprits were heatwaves (easily 'stopped' by the growing sales of air conditioners) and dengue fever, heavily dependent upon socio-economic factors, as demonstrated by a study which found much higher infection rates on the Mexican side of the border than in the contiguous US town of Laredo, despite mosquito vectors being more common in Laredo (Reiter *et al.*, 2003).

It is understandable why the ACF should sponsor research which ignores observational science in favour of questionable modelling results; it is after all an environmental pressure group. But why the AMA should put its name to the study is more difficult to comprehend, especially as (as the quotation at the beginning of this chapter indicates) received medical wisdom has long rejected climate alone as a significant factor in human health. Yet the World Health Organization sponsors reports claiming that 160 000 die every year because of global warming, and the *British Medical Journal* carries editorials suggesting that global warming will increase the number of deaths from extreme weather events (Patz, 2004), when observational data show that the number of deaths from such causes has actually *declined* during the period of recent warming. Indur M. Goklany, Assistant Director, Science and Technology Policy at the US Department of the Interior, very promptly pointed out in a 'Rapid Response' to the *BMJ* editorial that data on 'global annual mortality due to such disasters have declined from 73 700 in 1970–79 to 42 200 in 1995–2004 despite both a 50 per cent increase in population and increased warming of the world' (Goklany, 2004). And, as Goklany added, observational data indicate reductions in cold-related mortality outweigh increases in death from warmer weather in northern latitudes (Keatinge *et al.*, 2000). Similar observational research in the USA also failed to find an association between summer heat and mortality, again probably due to adaptation (Davis *et al.*, 2003).

Indur Goklany also wrote to the *Financial Times* on 1 February 2005 in response to reporting of a likely increase in weather-related natural disasters, pointing out that both the rates of death and absolute numbers of deaths from such events globally had declined markedly – 98.5 per cent for death rates, and 95.8 per cent for actual fatalities. Yet the linking of the power of nature to climate change is a temptation too great for some to resist, especially after the devastation Hurricane Katrina wreaked on New Orleans in 2005. Katrina was ultimately graded as a mere Category 3 hurricane, but it did what had long been feared, and scored a direct hit on New Orleans, which lay largely below sea level and suffered extensive damage and loss of life, largely from inundation by the storm surge. Hurricane frequency and intensity are known to vary according to cyclical patterns, but (while insured losses reflect a rising trend in the USA) the peer-reviewed literature (including the Third IPCC Report) does not suggest a link between climate change and long-term trends in hurricanes, droughts, wet spells, tornados or other severe weather. Yet, despite a letter published in *Science* on 9 December 2005 pointing this out (Pielke, Jr, 2005a), Donald Kennedy (2006), the editor of *Science* wrote in an editorial on 6 January 2006: 'The consequences of the past century's temperature increase are becoming dramatically apparent in the increased frequency of extreme weather events.'

The disjuncture between models and observational data also holds for the impact of climate change on biodiversity. There is evidence of species loss from particular habitats, but limited observational evidence (as we have seen) of actual documented extinctions over the past 100 years when there has been an equivalent warming. This is not to say that there are not many endangered species which warrant our concern, but we can only get to ecological Armageddon thanks to mathematical models, and they are misused by activists. For example, a study in *Nature* in January 2004 used computer models to simulate the way species' ranges 'are expected' to move in response to temperature and other climatic conditions under three different climate change scenarios. The study had an enormous error term, concluding that anything from 5.6 per cent to 78.6 per cent of species might be at risk – a range which was promptly ignored by environmental groups WWF and Friends of the Earth as they sought to exploit the research for fundraising, referring only to the possible one million species that might be made extinct. Sceptics were quick to point out that a similar warming observed over the previous years had not produced the extinction holocaust the models were predicting – except in runs of species–area models, but certainly not in observations. As Botkin *et al.* (2007) noted in suggesting ways in which forecasting of species extinctions due to global warming might be improved, during the geologically recent ice ages and warmings surprisingly few species *actually* became extinct.

Such modelled results reflect the prominence among scientists, particularly in the United States, of those who have been supporting a pessimistic view of environmental degradation since the re-emergence of Malthusianism from the late-1960s, exemplified particularly by Stanford University's Paul Ehrlich and his associates. Such scientists have been especially vigorous in defending their view of the world against challenges, and we examine in this chapter the behaviour of a group centred on Ehrlich in defending their worldview from a challenge mounted in 2001 by Bjorn Lomborg, a Danish statistician and self-confessed apostate former Greenpeace supporter. Ironically, Lomborg argued that there was a 'Litany' in environmental science that was not supported by the statistics; the reaction of Ehrlich and his supporters was worthy of any religious inquisition, and thus lent considerable weight to his claim of Litany. We will focus here on the politics of the science as a network of scientists centred on Ehrlich acted to defend the 'Litany' against the challenge Lomborg posed – thus proving his point.

EXCORIATING THE SCEPTICAL ENVIRONMENTALIST

In 1997, Bjorn Lomborg, an associate professor in the Department of Political Science at the University of Aarhus, in Denmark and a former Greenpeace supporter, started a study group among his students with the aim of taking apart the claims of US economist Julian Simon that the state of the world's environment, rather than going to Hell in a handcart, was actually improving. Simon had once famously bet with neo-Malthusian Paul Ehrlich that a specified basket of commodity prices would decline in real terms over a decade, thus giving lie to the pessimism of the Club of

Rome's *Limits to Growth* study which projected rapidly declining resources. Surely, if resources were becoming more scarce, prices would rise in response, Ehrlich wagered, but instead they fell. As an economist, Simon understood better the relationships between physical quantity, price, efficiency and technological innovation.

Ehrlich had made the claim in print that 'If I were a gambler, I would take even money that England will not exist in the year 2000.' Simon considered this too silly to bother with, but instead offered to wager that the real price of any commodity would be cheaper at any time in the future. Ehrlich and his colleagues (Berkeley physicists John Harte and John Holdren) could choose their own commodities and time period, and chose chromium, copper, nickel, tin and tungsten and a decade as the period. They bought a notional $200 worth of each, making a bet of $1000. Ehrlich and his associates lost badly, and mailed Simon a cheque for $576.07 – the amount by which the basket of commodities had become cheaper in real terms.[1] Simon was a notable contrarian, but had published widely in respectable outlets (Simon, 1996, for example). Despite this, Ehrlich dismissed him as a 'professor of mail-order marketing'.

The preliminary results from the study group, much to Lomborg's surprise, indicated that Simon had a point – that the state of the world was generally improving, according to the statistics published by governmental and intergovernmental organizations and other accepted sources. The preliminary results from the study group were presented in four articles in the Danish national newspaper *Politiken*, in 1998, and subsequently published by Lomborg in Danish as a book (*Verdens sande tilstand*). This book came under attack by environmentalists within Denmark, but this was a mere ripple when compared with the storm which erupted when Cambridge University Press published an extended and updated version under the title *The Skeptical Environmentalist: Measuring the Real State of the World* in September 2001 (Lomborg, 2001).

The book was favourably reviewed in *The Economist* and other newspapers, but a highly critical review soon appeared in the prestigious journal *Nature* (Pimm and Harvey, 2001). This was authored by Stuart Pimm of the Center for Environmental Research and Conservation, Columbia University, and Jeff Harvey of the Centre for Terrestrial Ecology, Netherlands Institute of Ecology, Heteren, The Netherlands. The tone of the review was set by the opening line: 'The subtitle gives the book away. It rehashes books such as Ronald Bailey's (1995) *The True State of the Planet*'. Bailey's book is something of a contrarian pot-boiler, but both Bailey's and Lomborg's books take their title from Worldwatch's Lester Brown and his regular *State of the Planet* tomes, which are very definitely activist volumes. Regardless of the quality of Bailey's book, it is a much shorter and less detailed volume, and it was a gross calumny to suggest that Lomborg's was merely a 'rehash' of Bailey's. Next, they criticized a book they were not reviewing, complaining that 'the book's origin was a class he taught in 1997. The original Danish version appeared a mere year later – remarkably fast, given the delays of academic publishing.'

Had Pimm and Harvey been reviewing *Verdens sande tilstand*, they might have had a mild rhetorical point (although Lomborg has stated the idea for the study group started in *February* 1997, so the genesis of that book was not as short as might appear), although it might also conceivably have reflected Lomborg's industry. But the book Pimm and Harvey were reviewing was, as noted above, not simply a translation of *Verdens sande tilstand*, but an extended and updated version published three years later. When they did finally get around to the book they were reviewing, they dismissed it as reading 'like a compilation of term papers from one of those classes from hell where one has to fail all the students. It is a mass of poorly digested material, deeply flawed in its selection of examples and analysis.' And that was just the first paragraph.

Pimm and Harvey immediately moved to the technique of guilt by association, stating that 'Lomborg admires the late Julian Simon' and then continued to attack Simon's views, not Lomborg's – ignoring completely the point that Lomborg had been motivated, not by admiration of Simon, but by a desire to rebut him. They claimed that: 'No external references support the ensuing paragraphs justifying that "things are getting better". Quoting the primary literature troubled Simon, too.' This was criticism bordering on the bizarre. As Lomborg later pointed out:

> That there is [*sic*] no external references to support the 'things are getting better' is explained in endnote 14: 'This and the following claims are documented in the individual chapters below.' It seems quite strange that Pimm & Harvey should expect the 'things are getting better' to be documented right there and then, as it is the subject matter of the entire book. (www.lomborg.com/critique, accessed 3 August, 2005)

The hostility Pimm and Harvey obviously held for the arguments of the book was clearly obscuring their reading of it.

The bizarre nature of the review continued. They criticized Lomborg for a 'bias towards non-peer-reviewed material over internationally reputable journals [that] is sometimes incredible.' In particular, they thought it poor that only 1 per cent of Lomborg's sources were original papers in *Nature*. But, as Lomborg noted:

> why should my book have referenced particularly *Nature* articles more? Are *Science* articles not as good? And what about articles from the multitude of

> other, more specialist journals – *Journal of the American Medical Association, American Economic Review, Papers and Proceedings, Environment, Energy Policy, Climatic Change* etc. just to name a few? (www.lomborg.com/critique, accessed 3 August, 2005)

The demand that Lomborg should have made more extensive use of scientific journals was a curious one, given the task he had set himself. He set out to show Simon was wrong (hardly out of a sense of admiration) and Simon had stressed that he only used official statistics, to which everyone has access and thus would be able to check his claims. As a statistician, Lomborg took this as a challenge, and was quite explicit in stating what his sources were and why he was using them. On page 31 of *The Skeptical Environmentalist* he states (and it is worth quoting the passage at some length):

> But for me the most important thing is that there is no doubt about the credibility of my sources. For this reason most of the statistics I use come from official sources, which are widely accepted by the majority of people involved in the environment debate. This includes our foremost global organization, the United Nations, and all its subsidiary organizations: the FAO (food), the WHO (health), the UNDP (development) and the UNEP (environment). Furthermore, I use figures published by international organizations such as the World Bank and the IMF, which primarily collate economic indicators.
>
> Two organizations work to collect many of the available statistics; the World Resources Institute, together with the UNEP, the UNDP and the World Bank, publishes every other year an overview of many of the world's most important data. The Worldwatch Institute also prepares large amounts of statistical material every year. In many fields the American authorities gather information from all over the world, relating for example to the environment, energy, agriculture, resources and population. These include the EPA (environment), USDA (agriculture), USGS (geological survey) and the US Census Bureau. Finally, the OECD and EU often compile global and regional figures which will also be used here. As for national statistics, I attempt to use figures from the relevant countries' ministries and other public authorities.
>
> Just because figures come from the UNEP does not of course mean that they are free from errors – these figures will often come from other publications which are less 'official' in nature. It is therefore still possible to be critical of the sources of these data, but one does not need to worry to the same degree about the extent to which I simply present some selected results which are extremely debatable and which deviate from generally accepted knowledge. At the same time, focusing on official sources also means that I avoid one of the big problems of the Internet, i.e. that on this highly decentralized network you can find *practically anything*.
>
> So when you are reading this book and you find yourself thinking 'That can't be true,' it is important to remember that the statistical material I present is usually identical to that used by the WWF, Greenpeace and the Worldwatch Institute. People often ask where the figures used by 'the others' are, but there *are* no other figures. The figures used in this book are the official figures everybody uses.

Given this quite clear statement, it is exceptionally unfair for Pimm and Harvey to have failed Lomborg on a test he at no stage set himself – and patently partial not to do the same for the Worldwatch Institute's *State of the Planet*. And, significantly, Lomborg was dealing with official statistics regarding the environment – clearly very much within his expertise as a statistician.

The criticism of Lomborg's references was all the more remarkable when we compare it with Ehrlich's most notable book, *The Population Bomb*. This entire book, co-published by the environmental interest group, the Sierra Club, contains but 55 footnotes, six of which contain interpretative discussion, four are self-referential, eight are to literature which does not appear to be peer-reviewed, one is to the work of a politician, and four are to speeches. One speech, the authority for a view on the fertility practices of Colombian women, was delivered by a colleague to Ehrlich's local Kiwanis Club at Palo Alto. This was at the least, a case of double standards.

Remarkably for an attack on the quality of Lomborg's sources, Pimm and Harvey resorted to anecdotal evidence to rebut one statement Lomborg had made, and it revealed that there had been communication among Lomborg's critics. They cited a passage from Lomborg stating that 'Scientific luminaries such as Harvard biologist E.O. Wilson and Stanford biologist Paul Ehrlich are the enthusiastic supporters of an ambitious plan . . . to move the entire population of the US . . . [which] would live in small enclosed city islands.' Pimm and Harvey stated that 'The reference is directly attributable neither to Wilson nor to Ehrlich. "Is it true?" we asked them. Ehrlich: "I know of no such plan. If there were one, I wouldn't support it." Wilson concurred.' (We shall return to this contact between Pimm and Harvey and Ehrlich and Wilson later.) Yet Lomborg had cited a news story in *Science* – not a peer-reviewed paper, but then such would be highly unlikely to contain such a statement of support. The reference to the 'ambitious plan' was a news article in *Science* (Mann and Plummer, 260 (1993), 1868–71) on the Wildlands Project (which boasts Ehrlich's endorsement on its website):[2]

> [T]he principles behind the Wildlands Project have garnered endorsements from such scientific luminaries as Edward O. Wilson of Harvard, Paul Ehrlich of Stanford (who describes himself as an 'enthusiastic supporter'), and Michael Soulé of the University of California, Santa Cruz, who is one of the project's founders. (p.1868)

Ehrlich also reviewed Lomborg's book and repeated the claim:

> The book is full of distortions, and demolitions of straw men, often 'documented' by repeated references to dubious secondary sources. Ed Wilson and I

> are 'enthusiastic supporters of an ambitious plan, the Wildlands Project, to move the entire population of the US so as to recreate a natural wilderness in most of the North American continent.' We do not support such a 'plan'; it does not exist. (2001)

Interestingly, the claim attributed to Lomborg appears in quotation marks in both the review by Pimm and Harvey and that by Ehrlich, but it is not a direct quotation from Lomborg, and substantially misrepresents what he actually wrote. Lomborg states not that 'the entire population' would need to be moved, but uses the following words: 'Inevitably, the implementation of such a scheme would involve mass movements of people' (p. 257). Pimm and Harvey and Ehrlich commit the same error in making Lomborg's argument seem more extreme, and the false denials of Wilson and Ehrlich more plausible.

Pimm and Harvey also resorted to the tactic of likening Lomborg to a Holocaust denier in pointing to the virtual nature of most of the species supposedly becoming extinct annually:

> The text employs the strategy of those who, for example, argue that gay men aren't dying of AIDS, that Jews weren't singled out by the Nazis for extermination, and so on. 'Name those who have died!' demands a hypothetical critic, who then scorns the discrepancy between those few we know by name and the unnamed millions we infer.

This is a fallacious argument. While any individual would be hard-pressed to name more than a few Holocaust victims, the identities of the overwhelming majority of them *are* known, or were known by those who survived. They had lives, families, birth records, bank accounts, friends, and so on. There is copious evidence that they existed and that they suffered at the hands of the Nazis. With claims by Norman Myers or Edward Wilson that 40 000 species supposedly become extinct every year, we have no strong evidence that they exist, or that they have ever existed, or ceased to exist, outside a mathematical model relating species and area.

What was more disconcerting was that IPCC Chairman Rajendra Pachauri later likened Bjorn Lomborg to Adoph Hitler in the Danish newspaper *Jyllandsposten* on 21 April 2004. Harvey and Pimm also demonstrated what is known in relation to internet discussion groups as Godwin's Law of Nazi analogies – that any argument will eventually get around to somebody accusing another of being like Hitler or the Nazis. (In some versions, the individual making the comparison is deemed to have lost the argument at that point, but it is accurately stated as: 'As an online discussion grows longer, the probability of a comparison involving Nazis or Hitler approaches.')[1]

Some immediately leapt to Lomborg's defence, with letters to *Nature*. Anthony Trewavas (2001) (correctly) saw Pimm and Harvey as defenders of neo-Malthusians like Ehrlich:

> Pimm and Harvey state that there are ecological laws that ensure the correctness of doom-laden predictions. Presumably one of these laws enabled the environmentalist Paul Ehrlich to state in his 1968 book *The Population Bomb*: 'The battle to feed all of humanity is over. In the 1970s and 80s, hundreds of millions of people will starve to death.' (letter to *Nature*)

Stephen Budiansky (2002) noted that Pimm had in 1998 told a meeting at the American Museum of Natural History that the world population might reach 40 billion by the end of the twenty-first century, whereas population growth had slowed and UN agencies estimate that the world population will likely level off at about 9 billion by about 2050. Budiansky found the review more revealing of its authors than of Lomborg:

> Lomborg's whole point is that the refusal of some environmental activists to deal honestly with the data harms the credibility of both environmental science and environmentalism. Pimm and Harvey in their review appear to have provided a further example in support of this thesis. (letter to *Nature*)

There were other critical reviews of Lomborg's book, but few quite as egregious as that by Pimm and Harvey published in *Nature*. *Science*, for example, published a review by Michael Grubb, formerly of WFF but now an academic policy researcher. Grubb summarized his assessment of Lomborg's book thus:

> Lomborg has compiled an immense amount of data to support his fundamental assertion that in many respects the environment is getting better rather than worse and to argue that we should not worry much about the state of the world. These are two distinct theses. For the most part, I find his analysis of the first contention compelling but his case for the second woefully inadequate. (2001, p. 1285)

In contrast to Pimm and Harvey, both biological scientists, Grubb found himself in general agreement with Lomborg's analysis, but was critical of his optimism that technological innovation rather than policy would ensure the improvement would continue. This in itself is a point that is open to contestation and must be settled on the evidence. Most would argue for a role for both technology *and* policy, but policy can be wrongly attributed with success, and (worse still) it can also be counterproductive. Unfortunately for Grubb, there is evidence of both these in the very example of Lomborg's he chose to make his point – air pollution, particularly in London.

Grubb thought Lomborg too dismissive of the power of policy when he stated:

> Air pollution in London has declined since the late nineteenth century, but for the greater part of the twentieth century this has been due to a change in infrastructure and fuel use and only slightly, if at all, connected to environmental worries expressed in concrete policy changes.

Grubb maintained that this unreferenced example was 'simply wrong'. But then Grubb (2001, p. 1286) himself fails to give a reference for his assessment that:

> The huge improvements in London's air have been very much driven by policy. Most radically, the 1956 Clean Air Act banned raw coal combustion across large swaths of London, and a long series of domestic and European legislation governing vehicle exhausts has done much to clean up mobile sources. The dramatic impact evident from 1957 onwards is obvious in Lomborg's graph.

Unfortunately for Grubb's argument, there *is* evidence that the success of the British Clean Air Act was illusory, and that socio-economic factors were more important. Thirty years previously, Howard A. Scarrow (1972) published a paper showing that socio-economic factors (greater affluence, housing improvement, associated fuel-switching from coal to gas and electricity) were all proceeding apace when the Act was adopted, and the success of the Act was largely illusory. Other analyses supported this view (Auliciems and Burton, 1973). This is a basic problem in policy evaluation, and the erroneous attribution of changed outcomes to a policy measure is a constant trap for policy researchers, as wrongful attribution can lead to the unwise adoption of similar measures elsewhere, or the continuation of poor policy. One classic example of wrongful attribution is the belief that the introduction of compulsory seat belt legislation in the Australian state of Victoria resulted in a significant drop in road deaths, whereas the same reduction occurred in other jurisdictions with no legislation as a result of less car travel, itself a result of the 1973 oil crisis (Adams, 1995). Regardless of the merits of compulsory seat belt legislation, it spread as the result of an illusory reputation for efficacy.

Air pollution policy also provides us with an illustrative example of the other pitfall of policy: counterproductive policy. Such is the nature of social phenomena that the consequences of any action are less predictable than is the case in the natural sciences. Individuals exhibit learning behaviour, take evasive action, or even actively resist the implementation of policies. But even when none of these obtain, policies can produce unintended consequences which might make the costs of a policy greater than its benefits, or even produce effects opposite to those intended. One notorious example of this was the US Clean Air Acts 1970, which created an incentive structure which meant older, more polluting plant was kept in commission longer, and air quality was judged worse than it would have been in the absence of policy intervention – if technological improvement and plant retirement had been allowed to simply run their course (Maloney and Brady 1988).

As a political scientist, the author has a disciplinary predisposition to accept the proposition that policy measures affect the rate of technological change, either by means of direct regulation or by altering price signals through taxes, depreciation allowances, subsidies, and so on. But the matter is at least more complex than Grubb was suggesting. Lomborg took issue with Grubb's review nevertheless, writing a letter to *Science* disputing Grubb's claim (2001, pp. 1285–6) that he only dealt with this issue in a single paragraph in the introductory chapter and without any references ('through 352 pages of text and 182 pages of footnotes, only one paragraph and one note (without a reference)'). Lomborg pointed out that his book dealt with this issue in the chapter on air pollution and, furthermore, he cited several studies that have found little or no impact from regulation. On the basis of this evidence, Lomborg concluded that 'while pollution has, of course, fallen, the difference between the rate of fall before and after 1956, or the difference between cities that did or did not have pollution plans, is not discernible' (p. 170). This was precisely Scarrow's finding 30 years earlier.

In writing to *Science*, Lomborg complained of another gross error in Grubb's review. Grubb wrote that in his coverage of official surveys of the state of the environment, Lomborg had neglected that of the European Environment Agency:[4] 'The European Union's official assessment is not even in the reference list.' Wrong, wrote Lomborg: it was there, at page 453 (European Environment Agency, 1999). Perhaps most important was what Lomborg saw as Grubb's complete misrepresentation of his argument, evident even in the title of his review (and Pimm and Harvey's) as one that no policy intervention on the environment was necessary. Rather, Lomborg's argument was that it was important to prioritize in making decisions allocating scarce resources to problems, and that misleading information resulted in wasteful, inefficient resource allocation. Despite these substantial errors in Grubb's review, four months after it was submitted, *Science* finally declined to publish Lomborg's letter. Christine Pearce, Associate Letters Editor, reponding to Lomborg wrote on 14 March:

> Your letter was forwarded to appropriate staff for their information, but only in cases of egregious misinterpretation or other extenuating circumstances do we

> think it warranted to publish responses from book authors. (www.lomborg.com/critique, accessed 3 August, 2005)

Significantly, however, Grubb largely sided with Lomborg on the basic statistical message that things were improving. The point for Grubb was really that 'The Litany' which Lomborg sought to criticize was something of a 'straw man' argument – that it was held these days by only a few individuals, and not really worth arguing with. Grubb wrote that:

> To any modern professional, it is no news at all that the 1972 *Limits to Growth* study was mostly wrong or that Paul Ehrlich and Lester Brown have perennially exaggerated the problems of food supply. (2001, p. 1285)

Pimm and Harvey demurred, it seems, and they were determined to contest vigorously the very statistics which Grubb accepted as being relatively uncontroversial. They were not alone, and they knew it, stating in their review that an 'industry' had arisen to 'debunk' the book. At that stage, they reported, it included a website, a series of essays planned for *Scientific American*, a guide for journalists being prepared by the Union of Concerned Scientists, and various published pamphlets. Interestingly, for a review in such a prestigious journal as *Nature*, the website it provided was www.antilomborg.com, and many of the commentaries on it came from an individual who had thrown a cream pie in Lomborg's face in an Oxford bookshop. It is unclear what Pimm and Harvey thought the existence of such as website would prove about the quality of Lomborg's book, but clearly they (and the editors) thought it worth including for the edification of their readers. In doing so the editors of *Nature* were making their prestigious journal into something resembling a political pamphlet.

This announcement of an 'industry' was a strange statement to make in a review in a scientific journal, but it was interesting mostly for what it revealed about communication between the reviewers and those participating in the 'industry' emerging to defend what Lomborg had called 'The Litany'. As we have already seen (above) Pimm and Harvey had already been in contact with Ehrlich and Wilson. Those contributing to the 'series of essays planned for *Scientific American*', the activities of the Union of Concerned Scientists and some of the 'pamphlets' (as we shall see) came from a relatively narrow group of largely self-interested scientists connected to Paul Ehrlich and eager that Lomborg's heresy should not go unpunished. Unlike Michael Grubb, who as a committed environmentalist largely had accepted that many of Lomborg's targets were exaggerations, and had moved on, these were scientists who had a substantial stake in these 'exaggerations' persisting.

The criticisms produced by the Union of Concerned Scientists, posted on their website and written between 6 December and 10 December came from: Dr Peter H. Gleick (Pacific Institute for Studies in Development, Environment and Security in Oakland California); Edward O. Wilson (Harvard University); Thomas E. Lovejoy (World Bank); Norman Myers (independent scientist); Jeffrey Harvey (again); and Stuart Pimm (again). Myers and Wilson joined Stephen H. Schneider and Lester R. Brown of the Worldwatch Institute (and a number of other WRI staff and others) on 12 December in a feature in the e-magazine *Grist*. *Grist* Assistant Editor Kathryn Schulz argued that Lomborg's 'real goal' was to divide the Left and discredit the environmental movement. The essayists in *Scientific American* included Schneider, Lovejoy and John Holdren. Not only were these critics the principal 'litanists' whose reputations Lomborg had called into question, they were a very small and tightly-defined group. They all seemed to be connected by an association with one person: Paul Ehrlich, who had famously lost the wager with the Julian Simon, the contrarian whose statistics Lomborg had set out to disprove.

Schneider had previously joined with Ehrlich to 'accept' a second bet with Simon (Ehrlich and Schneider, 1995). Simon declined because he had issued the challenge that any indicator of human welfare would improve over the next decade. Rather than taking an obvious indicator such as life expectancy or GDP per capita, Ehrlich and Schneider chose a number of environmental indicators, which Simon dismissed as having only indirect relevance for human welfare. (He pointed this out by suggesting that he had effectively been prepared to wager that the performance of athletes at the Olympics would improve with time, whereas Ehrlich and Schneider were wanting to bet on the track conditions.) Pimm had also previously criticized Simon on biodiversity (Pimm *et al.*, 1995). To Ehrlich and his supporters, it must have seemed as if Lomborg was Julian Simon reincarnate.

Besides this close tie with his Stanford colleague in Schneider, with whom he had also published,[3] Ehrlich had published with nearly all the Lomborg critics: Myers;[4] Wilson;[5] Holdren;[6] Pimm.[7] Moreover, Ehrlich had published with two other people who were very important in the science establishment in 2002: *Science* editor Donald Kennedy[8] and American Association for the Advancement of Science President Peter H. Raven,[9] both of whom had been colleagues at Stanford. Ehrlich had even introduced Kennedy to readers of *Science* upon his appointment as editor (Ehrlich, 2000a). But Kennedy, despite his association with Ehrlich in the nuclear winter alarm, oversaw a much more scholarly review of Lomborg's book than *Nature* or *Scientific American*, even though it was written by someone with a close association with the WWF. But the positions held by Kennedy and Raven indicated that Ehrlich's associates now controlled the commanding heights of science politics in the USA. It does not take a social network analysis, as Wegman used with the 'Hockey Team', to see that Lomborg's critics all had close associations with Ehrlich.

What was remarkable behaviour from Raven, as then current President of the American Association for the Advancement of Science, was that he and 11 others (including Edward O. Wilson) wrote a letter to Cambridge University Press in July 2002 essentially asking them to burn Lomborg's book. In their approach to CUP the 12 called Lomborg's book a 'non-peer-reviewed work' (*Time Magazine*, 2 September 2002, p. 58). They called on CUP to convene a scientific panel to identify every error and misrepresentation and add an errata sheet to every copy of the book, to transfer its rights in the book to a popular, non-scholarly publishing house, and to review its internal procedures to establish how CUP could have let through a book that was 'essentially a political tract' (Harrison, 2004). Despite being told that it had been peer-reviewed by four appropriate referees, Raven and associates continued to repeat the erroneous claim that it had not been subject to peer review, most notably in *Time* magazine on the eve of the World Summit on Sustainable Development in Johannesburg in September 2002 (Goldstein, 2002).

Even those who had not published with Ehrlich were close to him. For example, Peter Gleick was President of the Pacific Institute for Studies in Development, Environment and Security. His governing board was chaired by Ehrlich's wife, Dr Anne H. Ehrlich, and its advisory board included Dr Stephen H. Schenider and Dr Peter H. Raven. Ehrlich had written the preface for a book by Gleick (Ehrlich, 1998). All the fierce critics were closely associated with Ehrlich, and the attacks were quite clearly coordinated. Lovejoy, Myers, Wilson, Harvey and Pimm all contributed to the Union of Concerned Scientists counterattack. Stephen Schneider contributed to the *Scientific American* response and (with Myers and Wilson) to the environmentalist 'e-zine' *Grist*. Schneider also has connections to the UCS, having acted as its representative at at least one Conference of the Parties to the FCCC. One can but speculate about what communications passed among this scientific 'posse', but we know that Pimm and Harvey were in touch with Wilson and Paul Ehrlich, because they stated so, and Ehlich makes exactly the same misquotation of Lomborg as made by Harvey and Pimm over the Wildlands Project.

But it does seem that Lomborg was excoriated largely because he became intellectual heir to Simon, and Ehrlich and his supporters appeared determined to drive a stake through the heart of his credibility. Ehrlich seems to have been at the centre of the 'posse', even though he did not lead it – at least in the pages of *Nature* and *Scientific American*. Ehrlich (together with his wife) had also supported Lester Brown in an opinion piece in the

Washington Post (Ehrlich and Ehrlich, 1989). Of the authors of critiques of Lomborg in *Nature*, *Scientific American*, *Grist* and the Union of Concerned Scientists, seven (Brown, Gleick, Holdren, Schneider, Myers, Pimm, Wilson) had either published jointly with Ehrlich or had been supported by him in some public way. Three of the others were not scholars, but employees of environmental activist groups (World Resources Institute and Alliance to Save Energy). This was a posse formed to defend not just the Litany, but the reputation of Paul Ehrlich.

Ehrlich himself had begun from the outset to cultivate the myth about Lomborg's book that was a substantial slur: that it had not been subjected to peer review. He stated that:

> Cambridge University Press (CUP) obviously undertook no serious scientific review of the *TSE* manuscript and printed a decomposers' dream that it claims explodes 'the widely propagated [*sic*] myth that the state of the environment continues to spiral downwards beyond our control.' CUP should be ashamed of abandoning academic standards and should be worried about whether competent scientists will now publish with them. It is supporting powerful economic interests that are anxious to convince us that business as usual is not wrecking human life-support systems. (Ehrlich, 2001, p. 51)

As late as October 2003, Ehrlich was continuing to smear the book in this fashion:

> The book gained much momentum by being published by Cambridge University Press. That gave it the aura of peer-reviewed work, even though it clearly had not been vetted by any environmental scientist. It eventually did receive peer review – a large number of uniformly scathing analyses in scientific journals. (Ehrlich, 2003)

There were few analyses that were 'uniformly scathing' beyond the environmental movement and Ehrlich's circle of associates. Lomborg immediately contested this claim in a letter to the editor, revealing that Ehrlich had written to Cambridge University Press making such a claim and urging them to stop publication. Cambridge informed Ehrlich that he was wrong, and that the book had been reviewed by four distinguished scientists, including environmental scientists. Ehrlich was repeating a falsehood he had been told two years previously was a falsehood. (In the same opinion piece he again labelled Julian Simon a 'professor of mail-order marketing' and again claimed (wrongly) that he was Lomborg's 'hero'.)

Schneider (*Scientific American* p. 65) also suggested that peer review by natural scientists did not occur: 'that the natural scientists weren't asked is a serious omission for a respectable publisher such as Cambridge University Press'. Schoenbrod and Wilson (2003, pp. 599–600) took the trouble to check with Cambridge editor Chris Harrison who reported that the book was reviewed by two referees from environmental science departments, one from climate science and one social scientist, and all recommended publication. Twelve scientists led by American Association for the Advancement of Science president, Peter Raven, in an approach to CUP also called Lomborg's book a 'non-peer-reviewed work' (*Time*, 2 September 2002, p. 58). This claim was repudiated by CUP. According to editor Chris Harrison (2003), the reviewers were from the UK and the USA, and three of them were already reviewers at the Press. Of these three, one was a climate scientist, the second an expert in the economic consequences of climate change, and the third an expert in 'biodiversity and sustainable development'. Harrison expected in advance that the statements from these reviewers would be mixed, and that in the end the conclusion would be not to publish the book. But to his surprise, all four recommended publishing.

Ehrlich – perhaps mindful of his spat with Simon – dismissed Lomborg's research as 'an old story', and he claimed that Lomborg's book was 'being very heavily promoted for political purposes' (*San Francisco Chronicle*, 4 March 2002). Politics certainly played a part in the remarkable series of review articles then published in *Scientific American*, because this series was published after lobbying by a number of scientists. Editor John Rennie admitted that the initial intention was to simply publish a standard review of Lomborg's book but 'we started to hear very clearly from scientists [saying] the book was doing a disservice to their field' (*San Francisco Chronicle*, 4 March 2002). Rennie also took a deliberate decision to run the attack pieces without offering Lomborg a right of reply, arguing that 'we felt it would not be a terrible disservice' to allow Lomborg a reply in a future issue. A reply from Lomborg was 'tentatively scheduled' to run in May, but by this time the posse would have had an uncontested run in the media of four months – a substantial disservice to notions of fairness to say the least.

What then happened was also remarkable. Denied a right of reply, Lomborg answered his critics on his website, reprinting the *Scientific American* critiques and answering them paragraph by paragraph. This was met by a demand from *Scientific American* that he remove their material from the site on the grounds that posting it breached copyright. Patrick Moore, an apostate founder of Greenpeace, then published the material on his *Greenspirit* website – and challenged *Scientific American* to sue.

Rennie himself responded in print to Lomborg's reply, confirming that the thrust of the critiques published previously accorded very much with the views of the editor. He took it upon himself to reply, he stated, because most of the original posse were 'unfortunately not available' (Rennie, 2002). Rennie repeated the myth that Ehrlich began and continued to perpetuate: that *The Skeptical Environmentalist* could not have passed peer review, and therefore that it had not been peer reviewed. He also confirmed that the reviewers of Lomborg's book were not chosen randomly, but because they were 'leading spokespeople for the science that Lomborg criticizes'. They were selected, in other words, not because they were leading scientists, but because they were leading *spokespeople for science*.

Rennie's piece was full of editorializing. Lomborg was accused of making sly comments ('one of Lomborg's sliest comments'); 'Lomborg sniffs', rather than writes. As noted elsewhere, Lomborg's primary target was what he termed 'The Litany' – the bad news story promulgated by NGOs and science activists which he argued was not supported by the data. His data was overwhelmingly drawn from various official statistics, rather than primary research papers. He was criticized for not relying on scientific articles – a criticism which missed the point totally. Yet he was also criticized for accepting The Litany as being an accurate representation of the underlying science, as if it was *his* responsibility (rather than that of those who repeated The Litany) to be accurate. Moreover, what was notable was that the scientists who excoriated Lomborg had been largely silent on the misrepresentation of science by the Litanists.

For example, Rennie criticized Lomborg for citing Norman Myers's 1979 estimate of annual species extinction rates of 40 000. Rennie stated that these were early attempts to quantify extinction rates, and 'they look crude in retrospect'. Researchers have revised these estimates, wrote Rennie, but 'those figures live on in some popular writings and they should not, but they are also not the figures that serious scientists still use in their work.' Perhaps not, but that missed Lomborg's point: these figures and even more exaggerated figures *are* used by Litanists, and are used without fear of contradiction by Myers, Lovejoy, Pimm and Wilson, and without scientists lobbying *Scientific American* to publish critiques because their science is being distorted by 'egregious distortions'. For example, as we noted earlier, a Greenpeace advertisement soliciting donations (and which appeared at about the time Lomborg was writing) claims a rate of 50 000–100 000 extinctions annually. (A similar charge can be levelled against the officials of the IPCC, who have been known to issue press statements rejecting dissident viewpoints, but remain silent over misrepresentations by supportive NGOs). In fact, as Rennie acknowledged, Lomborg and Lovejoy were pretty much in agreement on estimated extinction rates; they differed on what these numbers meant – the objection was really to how Lomborg 'framed' the problem, and the value he attached to biodiversity.

Actually, Lomborg suggested a rate of extinction 1500 times background, while Lovejoy replied with a *lower* estimate of 1000 times background in his criticism, providing evidence for one analyst that 'biodiversity illustrates the point that experts tend to interpret ambiguous data in a way that supports their prior expectation' (Balint, 2003, p. 17). Rennie in fact stated that it was Lomborg's arguments that were at issue, not the correctness of his statements. As Balint put it, the differences were 'primarily ones of interpretation, emphasis and synthesis, rather than of hypothesis formulation, data collection, and experimental result'.

That the objection to Lomborg on biodiversity and species loss centred on how he constructed the science is interesting, because it points to the possibility that this was the basis of most of his critics' objections. And in criticizing him for attacking straw men, rather than the best science has to offer, they inadvertently granted him victory in his argument, because his main theme was that the Litany is not supported by the science. And the Litany had been constructed with the tacit consent of scientists such as those who formed the posse, because they had allowed the 'egregious distortions' to stand in the popular press and the political realm – because, one presumes, the Litany supported policy prescriptions of which they approved.

An analysis of Lomborg and his critics confirms this suspicion: for all the words written in the reviews in *Nature*, *Scientific American*, and *Science*, his critics barely landed a substantive blow on Lomborg. Lomborg made a few factual errors, which he acknowledged, that were of a minor or inconsequential nature. At one stage he referred to 'energy' rather than 'electricity' and in another he referred incorrectly to 'catalysis' rather than 'hydrolysis'. (The latter appears to be a translation error, as it did not occur in the Danish edition.) Legal scholars Schoenbrod and Wilson (2003) examined the charges levelled against Lomborg as if preparing a brief for prosecution, and found that only 11 charges of factual error were made in the several pages of the *Scientific American* critique, and only these two had any substance.

Most of the charges levelled against Lomborg fell into three other categories: personal attacks; supposed straw man arguments; and arguments that he had framed the facts incorrectly. Schoenbrod and Wilson noted (2003, p. 588) that 'The first category – personal attacks – has a surprisingly large number of entries given that science is supposed to be a search for truth.' The straw man category had a few entries, and the most numerous category of charge levelled was that Lomborg had framed the facts inappropriately. The result of this forensic analysis was ironic, given that Lomborg was then subjected to legal inquisition for 'scientific dishonesty' in his native Denmark.

LOMBORG ON TRIAL

The Lomborg case soon took a new turn in the politics of science, for after the election of a conservative government in Denmark, Lomborg was appointed Director of the Environmental Assessment Institute (on 26 February 2002). In anticipation of this appointment, biologist Kåre Fog, of the Danish Ecological Council, lodged a complaint on 21 February with the Danish Committees on Scientific Dishonesty (DCSD) over Lomborg's book. His motivation was quite open:

> The reason for submitting this complaint at this time is that, at the time when the board of the Institute is to elect who is to be employed, the board should be clear that there are doubts regarding Bjørn Lomborg's scientific integrity. (Fog, undated Internet page)

Fog was thus dissatisfied with the prospect of the appointment of Lomborg to a position of authority by the newly-elected Danish government. Fog's complaint was joined on 7 March by one from Mette Hertz and her husband Henrik Stensdahl, the latter the technical director of the Danish wind energy company, Bonus Energy, beneficiary of both Danish government subsidies for wind power development and the beneficial effect of climate change policy on the renewables industry as a whole. Finally, on 22 March a further complaint was lodged by Fog on behalf of Stuart Pimm and Jeffrey Harvey, Lomborg's reviewers in *Nature*.

The DCSD, established to investigate cases of scientific fraud and malpractice, released its findings on 7 January 2003, declaring that the book fell within the concept of 'objective scientific dishonesty' and was 'clearly contrary to the standards of good scientific practice'. To reach this finding, the DCSD had to engage in some tortured logic, declining to rule on the question as to whether the book was a work of science rather than a 'debate-generating book': to accept it as science would have enhanced its standing in the eyes of some; to deny that it was science would have removed it from its jurisdiction. Remarkably, the DCSD relied almost entirely for evidence of misconduct upon the reviews in *Scientific American*, themselves not peer-reviewed science. And the DCSD stated quite explicitly that it was motivated by the political impact the book might have, pointing to favourable press in the USA which had the highest energy consumption in the world and 'powerful interests . . . bound up with increasing energy consumption'. As Roger Pielke Jr has pointed out, several of Lomborg's critics, including Peter Raven and Stuart Pimm were also concerned with the political uses to which the book might be put (Pielke, 2004, p. 408). The Union of Concerned Scientists had justified the offensive as a pre-emptive political strategy because 'this book is going to be misused terribly by interests opposed to a clean energy policy' (Pielke, 2004, p. 407).

There was a similar motivation for one of the *Nature* reviewers and complainant to the DCSD, Jeff Harvey. Harvey, once on the editorial staff at *Nature*, made this extraordinary statement on an Internet discussion site:

> It seems clear to me that the media is largely supporting Lomborg because his message, as scientifically fraudulent as it is, bolsters the arguments of those on the political right and lends weight to a corporate-driven political agenda; many of these transnationals are corporate sponsors and advertisers in the media, and one of the most hostile is that of the Murdoch empire, which owns the rabidly conservative anti-environmental Fox media chain in the US and the *Times*, *Sunday Times* and *Sun* group network in the UK. (Harvey, 2002)

Many academics in Denmark leapt to Lomborg's defence. Professor Peter Pagh complained to the Ombudsman about the DCSD ruling, which he found had 'serious legal omissions'. Seven professors of social sciences published an open letter to the DCSD protesting the ruling. On 18 January *Politiken* published a letter of protest against the Lomborg decision from the DCSD signed by 286 research scientists. Then, on 21 January the Ombudsman referred Professor Pagh's complaint to the Ministry of Science, Technology and Innovation – the ministry responsible for the DCSD. On 30 May, a working party of the Ministry of Science, Technology and Innovation published a report stating that the DCSD has exceeded its statutory authority in the Lomborg case, and then finally Lomborg was exonerated on 17 December 2003 when the Ministry of Science, Technology and Innovation published its final report on the DCSD's ruling regarding the Lomborg book.

The Ministry found that the DCSD judgment was not backed up by documentation, and was 'completely void of argumentation' for the claims of dishonesty and lack of good scientific practice. The Ministry characterized the DCSD's treatment of the case as 'unsatisfactory', 'deserving criticism' and 'emotional' and pointed out a number of significant errors. The DSCD's verdict was consequently remitted.[10]

The Report was damning. It found, *inter alia*:

- The DCSD has not substantiated its ruling. Dr Lomborg has not been told exactly where he has, allegedly, made mistakes. This is a case of *'significant neglect in case processing by the DCSD.'* *'Here the Ministry must point out that the DCSD has not documented where the respondent (BL) has allegedly been biased in his choice of data and in his argumentation, and that the ruling is completely void of argumentation for why the DCSD find that the complainants are right in their criticisms of BL's working methods. It is not sufficient that the criticisms of a researcher's working methods exist; the DCSD must consider the criticisms and take a position on whether or not the criticisms are justified, and why.'* (Point 6.1)

- In its ruling, the DCSD emphasizes that scientific work should go through 'peer review'. Nevertheless, the DCSD omits to examine whether this has happened in the case of Dr Lomborg's book. The Ministry describes this as '*dissatisfactory*' [*sic*]. Dr Lomborg's book was accepted at the Cambridge University Press after a thorough peer review by four recognized scientists. (Point 6.5)
- The people who have complained about Dr Lomborg have been incorrectly treated as 'parties' to the case. This means that they have been heard to an extent they have not been entitled to. This could have had the consequence, '*that the DCSD have attached too great an importance to the relevant complainants' assessments, and this could have meant that the time for case processing was extended as complainants were allowed a longer hearing than they were entitled to. The hearing in cases that do not include parties should also consider the interest of the respondent in having the case concluded.*' (Point 6.2)
- It was '*clearly wrong*' that Dr Lomborg was not heard before public disclosure of the DCSD ruling. (Points 6.6.2. and 6.8)
- Criticism of the fact that the chairman of the sub-committee in the case of Dr Lomborg came from the health sciences and not from the social sciences, which is Dr Lomborg's field. (Point 6.6.1)
- Since Dr Lomborg's book was published outside Denmark, it is doubtful if the DCSD has the competence to try the case. The DCSD ought to have checked this. (Point 5.1.5)
- It was a mistake, that in the judgement of Dr Lomborg's cost–benefit analyses two references were used, which have been released after the publication of Dr Lomborg's book. (Point 6.1)

One significant point about the hostile reviews and the proceedings taken by the DCSD is that they were not isolated events with separate acts of initiation. As noted above, there is evidence of contact between Pimm and Harvey and Ehrlich and Wilson because Pimm and Harvey stated they were in contact. But there is additional evidence that there was communication between Union of Concerned Scientists critics and those in *Scientific American*, because Mahlman referred to Stephen Schneider's 'in press' piece for *Scientific American* in his UCS post on 6 December 2001. Pimm was also in contact with a Lomborg opponent in Denmark, and they apparently also organized for Pimm to lodge a complaint with Aarhus University, where Lomborg worked. Pimm wrote on 14 January 2002 urging the University to investigate Lomborg's alleged errors. Shortly after, additional letters of complaint were sent by E.O. Wilson and by Tom Malone, General Secretary of the International Council of Scientific Unions (ICSU), and a senior scientist with involvment in international science agendas stretching back to

Global Atmospheric Research Program (GARP), International Geosphere Biosphere Program (IGBP), and Scientific Committee on Problems of the Environment (SCOPE) (Fog, n.d.).

The complaints to the DCSD were also coordinated in part. That from Henrik Stiesdal and his wife Mette Hertz on 7 March 2002 appears to have been independent, but that lodged by Stuart Pimm and Jeff Harvey on 22 March was undertaken in close cooperation with biologist Kåre Fog, the first complainant. Fog clearly shared Harvey's political views, stating in his account (Fog, n.d.) of the case that 'There exists no evidence that Lomborg is supported by the oil industry, but his actual actions are exactly what the oil industry might want him to do.' This was little short of scientific McCarthyism.

Clearly, the uses to which Lomborg's book might be put were of overwhelming importance to the Litanists, who were exercised to a degree that could not be justified by the publication of work that was simply critical of their work. In a particularly revealing letter, Paul Ehrlich demonstrated considerable hostility in response to a communication from Lomborg seeking clarification of his reference to the support offered by Ehrlich and Wilson for the Wildlands project (referred to above), which had been reported in a news story in *Science*. Jeff Harvey (referring to Ehrlich as 'Paul') dismissed this report as having been 'written by two noted anti-environmental writers' and cited the text of the reply Ehrlich sent to Lomborg on 28 November 2001 (Harvey 2002):

> Dear Mr. Lomborg,
>
> The Pimm/Harvey review, which was much too kind to your book, is absolutely correct. As with most of the other issues you covered, you misread or misinterpreted Mann and Plummer's mediocre article, failed to go to original sources, and generally bungled the story. I suggest you spend the next few years studying elementary environmental science, publish some papers on it in the refereed literature, and then join the debate in any area where you have achieved at least basic competence, if you can. You might also flunk the undergraduates who wrote your book.
>
> Paul R. Ehrlich

This somewhat excessive, clearly passionate and (frankly) immature concern with the possible political implications of a book impugning what Lomborg and Simon described as 'The Litany' clearly lay behind an attack that was at least in large part coordinated, rather than left to chance. It demonstrated what Roger Pielke Jr has described as a politicization of science by scientists, who went beyond a policy perspective (providing a range of alternatives for policy-makers to choose from) to take a political perspective (seeking to limit the range of alternatives open to policy-makers). Pielke saw the whole episode as typical of an increasing tendency for 'scientific' debate on environmental issues to 'morph' into political debates: 'In many instances science, particularly environmental science, has become little more than a mechanism of marketing competing political agendas, and scientists have become leading members of advertising campaigns' (Pielke Jr, 2004, p. 406).

Pielke attributes this phenomenon both generally and specifically in the case of Lomborg to the adherence of the scientists in question to a linear model of science's relationship to society – 'get the facts right, then act'. As Pielke (2004, p. 406) notes, the world is rarely that simple, and the actions of the scientific posse seeking to counter Lomborg betray a naïve view that it is – and an adherence to a particular political agenda or set of policy prescriptions that they think depend upon *their* view of the state of the planet prevailing. This points to an insecurity on their part: if they were confident that their view was not a 'Litany', but based on incontrovertible science, they would hardly have felt the need to bother to counter the heresy published by a Danish statistician in a political science department. The 'virtual' nature of much of the science under discussion *allows* a greater degree of subjective projection of values on to the conduct of science, but it might well be that the reliance of much of environmental science on 'virtual numbers' – proxies and model results – also provides the motivation for corruption in the name of the noble cause of environmental protection. As Waterton and Wynne argued in the context of the European Union:

> The virtues of quantification and objectification lie in their ability to provide a kind of abstraction from messy and deeply complex contextual factors – the factors which have come to be called 'human perversity' but which could equally be described as political, structural, cultural and institutional factors. (1996, p. 435)

Waterton and Wynne refer to Porter's (1995) 'counter-intuitive observation' that technocratic discourses of quantification and standardization, 'actually reflect a fundamental *weakness and lack of political–cultural legitimacy*' (original emphasis) rather than the excessive power of agents of technocratic political control (Waterton and Wynne, 1996, p. 436).

If we accept this line of reasoning, the Litanists thought that they enjoyed political influence beyond what they might reasonably expect *because* they were able to exercise a numerical hegemony through a set of numbers that contained – and thus smuggled into political discourse in the guise of science – a set of political values and assumptions that gave them a strategic advantage (supporting alternative energy agendas, endangered species legislation, etc.). Had they greater confidence in their 'science' such a vigorous defence would hardly have been needed. Of course, there *are* numerous arguments in favour of species conservation, confronting the climate change issue, and so on, but they require more than just science to justify them, and (by requiring a discussion of ethics, economics, politics, and so on) they destroy the monopoly that scientists enjoy under a linear model. As Pielke put it, 'For those with scientific expertise, it . . . makes perfect sense to wage political battles through science, because it necessarily confers to scientists a privileged position in the political debate' (Pielke, 2004, p. 410). Much of the criticism of Lomborg centred on his fitness to engage in 'science' of this kind, but his real fault was to move the focus of the debate from the data to the frame which surrounds it – the meaning and significance attached to it, which formed the basis of the overwhelming majority of the criticisms levelled at his book.

Lomborg's 'sin' was therefore to threaten the monopoly of the activist scientists exemplified best of all by Paul Ehrlich. Kysar and Saltzman (2003) suggest the controversy reflects a typical 'environmental tribalism'. As these authors (2003, p. 1100) noted, with rare exceptions, Lomborg's critics were all scientists and professional staff members of environmental organizations, while his supporters were largely social scientists. Pielke adds that the conflict over who should win in the war between the tribes 'is fundamentally about values and in democratic systems is resolved through bargaining, negotiation, and compromise, i.e. politics' (Pielke, 2004, p. 415). He may well be correct in adding that 'Science has become the weapon of choice in this war' but he misses an important point from all of this which is made by Naomi Oreskes in a paper which accompanies his in the same journal and which was presented alongside his at the AAS symposium: that Lomborg adopts an essentially anthropocentric position, and many of his critics adopt an ecocentric one (though, as we shall see, they are somewhat confused about this). Lomborg is concerned with human welfare, in other words, whereas his critics are concerned with the intrinsic value of ecosystems. As Oreskes (2004, p. 376) puts it in relation to Rachael Carson, 'Carsons' argument was about things that can't be counted, yet still count.' (Ironically, Lomborg's critics seem to want such things to be expressed numerically, as if that charges them with greater persuasive power.)

The position was exemplified (again) by Harvey (2002) on an Internet discussion site:

> The reason it is appalling . . . is that Lomborg . . . sees ecosystems simply as utilitarian sources, while dismissing (or clearly not understanding) the connection between ecosystem health and the prosperity of human society.

Leaving aside the confusion evident here over the fact that the 'nature's services' argument is in itself essentially one about 'ecosystems as utilitarian sources', this attitude (common to many of the critics, as Oreskes notes) is that it is essentially a moral position which places it outside the utilitarian, liberal discourse which permits the 'bargaining, negotiation, and compromise' Pielke expects to resolve such conflicts. Such environmentalism seeks solutions through *radical* regulatory politics, based on moral positions which are distributed among the population in a largely bi-modal fashion. Such politics is quite different from *liberal* regulatory politics, where preferences follow a largely normal, bell-shaped distribution and policy can be (constantly) adjusted by incremental compromise to resolve political conflicts.

CONCLUSION

This distinction between liberal regulatory politics and moral-based regulatory politics is important, and we shall return to it in the final chapter in discussing some issues in the philosophy of science and politics of science, but for now we can note that the reaction to Lomborg was far more spirited in the United States than it was in the United Kingdom (or almost anywhere else). A similar pattern was apparent in the previous chapter, with the controversy over the Hockey Stick being particularly politicized in the United States. Hans von Storch exemplified the tenor of European climate scientists – reflecting critically on noble cause corruption – but there were others such as Zorita, Bürger and Cubasch who continued to adhere to scientific terms of engagement. These scientists consider there is anthropogenic warming occurring, but neither this nor their political beliefs prevented them from finding problems with the Hockey Stick. This points to a greater politicization of science in the USA than elsewhere, and I shall examine the reasons for this in the next chapter.

The attack on Lomborg came from a small circle of critics, mostly in the USA and mostly closely associated with Paul Ehrlich. As we have seen, it had more to do with the way Lomborg framed the science rather than with the science itself. Lomborg even believed the rate of extinction to be higher than his critics, and his sin was to regard this as less serious then they did. They could establish few errors of fact, and Lomborg quickly admitted to the few they did, which were rather trivial to his argument. When the controversy spread to his native Denmark, it did so with political complaints to the DCSD, which were upheld only because of its indefensible mode of procedure, which broke almost every principle of natural justice. On both sides of the Atlantic, there was an overwhelming concern with the politcal implications of his findings, and the swarming of defenders to the infectious ideas presented by Lomborg simply proved his point even more persuasively than his book itself that there was a Litany which was defended on the basis of a very shaky scientific foundation.

We shall engage with the enhanced political nature of the scientific controversies in the USA in our cases in the next two chapters, suggesting that the explanation of this lies partly in an apparent greater partisanship surrounding science in the USA (which probably reflects a more decentralized distribution of political power) and a (possibly related) tendency for radical regulatory politics to occupy the political Left, probably due to the arrested development of more conventional Left positions.

NOTES

1. http://en.wikipedia.org/wiki/Wager_between_Julian_Simon_and_Paul_Ehrlich, accessed 11 November 2005.
2. www.wildlandsproject.org
3. Kennedy *et al.*, 1998.
4. Ehrlich *et al.*, 1997; Myers *et al.*, 1993; Ehrlich, 1984.
5. Bazzaz *et al.*, 1998; Ehrlich, 1995; Ehrlich and Wilson, 1991; Ehrlich, 1988.
6. Holdren, *et al.*, 1995; Ehrlich and Holdren, 1988; Ehrlich *et al.*, 1977; Ehrlich and Holdren, 1975; Holdren and Ehrlich, 1974; Ehrlich *et al.*, 1973.
7. Ehrlich *et al.*, 1994.
8. Kennedy *et al.*, 1998; Kennedy *et al.*, 1999; Ehrlich *et al.*, 1984.
9. Ehrlich, 2000; Bazazz *et al.*, 1998; Ehrlich and Ehrlich, 1993; Ehrlich *et al.*, 1983; Ehrlich and Raven, 1967; Ehrlich and Raven, 1969.
10. A summary in English of the Ministry's assessment was provided at www.imv.dk.

5. Sound science and political science

I despise exaggeration – 'taint American or scientific.

Rudyard Kipling

There was considerable irony in the exaggerated response to Lomborg's book, because it rather proved his claim that there was an exaggerated Litany, not supported by science, that the world was on course for unmitigated ecological catastrophe. It is uncertain whether the attack by Lomborg's critics was 'un-American', but it certainly wasn't very scientific, and was distinguished by the distinct absence of errors of substance it identified in Lomborg's book.

All areas of science are acknowledged to be subject to error, deliberate or inadvertent. There is perhaps most attention focused on the causes and prevention of error in medical science, where errors can be a matter of life or death, and there are numerous studies which have sought to identify sources of error.

For example, one study has suggested that medical research findings are less likely to be true under certain conditions, such as when the size of the effect being measured is smaller, where there is greater flexibility in research design, definitions, outcomes and analytical modes, and when more teams are involved in a competitive chase of statistical significance (Ioannidis, 2005). There is also a greater chance of false findings when there is greater financial interest involved; in other words, despite what is known as the genetic fallacy (that there is no logical reason why the origin of research should produce 'contaminated' findings), the presence of financial interests should alert us to the possibility that financial incentives might lead to errors which favour those interests. Despite this, the drawing of the connection with financial interests is not sufficient in itself as an argument (in the absence of reason and evidence) that certain findings can be dismissed. And financial interests are but one source of bias; other prejudices can be just as powerful. There is a strong ethical principle in medical research that all conflicts of interest be declared, but rarely do we find declarations of political conflict of interest in the broad field of what we might broadly call 'environmental science'. So Lord Robert May can simultaneously dismiss climate sceptics while not declaring his close association with an environmental lobby group, the Worldwide Fund for Nature.

With medical research, particularly that involving new medications, the stakes are high, as development costs are considerable and product approval is important to drug companies, so we are not surprised when the scientific process sometimes fails to detect mistakes. A study of the painkiller Vioxx in the *New England Journal of Medicine* in 2000 omitted to record that several patients had suffered heart attacks. A study on the painkiller Celebrex in the *Journal of the American Medical Association* was discredited after it was revealed the authors had submitted only six months of data rather than the 12 months they had collected. (Both drugs were later withdrawn from sale.) Dr John R. Darsee was found to have fabricated much of the data for more than 100 papers he wrote while working at Harvard and Emory Universities and published in leading journals, including *The New England Journal of Medicine*, *American Journal of Cardiology*, and *Proceedings of the National Academy of Sciences*. Eric Poehlman was given a prison sentence and banned from ever again receiving public research money in 2006 in a scientific fraud case that led to the retraction or correction of 10 scientific papers (Interlandi, 2006).

But other research has slipped through the quality assurance processes. In 1999, federal investigators found that a scientist at the Lawrence Berkeley Laboratory had faked what was hailed as crucial evidence linking powerlines to cancer. Research linking MMR vaccines to autism was later discredited, but caused widespread alarm and a drop in immunization rates as the result of parents' concerns. In June 2005 a survey of 3247 scientists by the University of Minnesota and HealthPartners Research Foundation reported that up to a third of the respondents confessed they had engaged in ethically questionable practices, from ignoring contradictory facts to falsifying data (Martinson *et al.*, 2005).

Medical science, however, continues to provide examples of good science, where orthodoxy is overturned to great benefit. One such example was the overturning of the 'stress' theory of gastric ulcers by two Australian medical researchers, who showed that a bacterial infection was usually responsible (and won themselves a Nobel Prize in the process). Another recent case is that of Ananth Karumanchi, whose research has led to a much improved understanding of pre-eclampsia, a condition thought to kill 75 000 women annually. Karumanchi's research, initially rejected by the journal *Nature*, had to overturn the erroneous assumption of biologists that the mother and the foetus had an underlying 'harmony of interests', whereas the cause lay in maternal–foetal conflict (Groopman, 2006).

Part of the problem is a conjunction between the competitive pressures under which scientists work (including, but not confined to, the 'publish or perish' imperative) and the sheer number of periodicals – over 54 000 are listed in *Ulrich's International Periodicals Directory*. With an explosion in the number of journals, it is thought that sheer numbers have overwhelmed quality assurance processes. But this does not appear to account for the problems with major journals, except that the recent and growing assessment of research quality in most universities has led to an explosion in submissions to leading journals. Most of those submitted are rejected before being sent out to review, and are therefore assessed by the editorial staff rather than by scientific peers. The role of editors has always been important, but it has increased in importance as the volume of submissions has grown, and the qualifications and expertise of editors is increasingly coming under scrutiny. Editors at *Nature* are relatively young and inexperienced: median age is mid-30s (mid-40s at *Science*). *Nature* has no editorial board. Along with increasing submissions to leading journals, there is more emphasis on networking to personalize the process as authors seek to gain an edge. The importance of rejection without review has thus increased, along with fears that reviewers sabotage papers that compete with their own (McCook, 2006).

This leads us to consider some important questions about the conduct of science, particularly how we can judge 'good' or 'sound' science and how politics might affect it. This is particularly important with environmental science, because one could easily gain the impression from political debates in the United States in particular, that how one views environmental science is a matter of partisan politics. One author captured this claim in the title of his book: *The Republican War on Science* (Mooney, 2005). The reaction to President George W. Bush's 'sound science initiative' saw actors on the Left of politics condemning the move to ensure regulatory decision-making was based on 'sound science' as decidedly partisan. It saw accusations that the Bush Administration was 'politicizing' science by demanding that science informing policy must meet certain criteria, such as publication after peer review. (The Left-leaning Union of Concerned Scientists led the charge against Bush with a petition in February 2004.) This seemed to fly in the face of Michael Crichton's statement that 'Data's not Democratic or Republican, it's data' (Crichton, 2005).

The partisan politics can be vigorous, to say the least. John H. Marburger III, Bush's science adviser, a lifelong Democrat, claims that the complaints about the 'attack on science' were a distortion (Smith, 2005). Marburger dismissed the UCS petition, and was promptly called a 'prostitute' by Harvard psychologist Howard Gardner. Many of the scientists openly endorsed John Kerry in the 2004 campaign, including climate scientist James Hansen, who campaigned for Kerry in his home state, the key state of Iowa. The matter continued: in late June 2005 Senator Richard J. Durbin (Dem, Ill.) introduced the Restore Scientific Integrity in Federal Research and Policymaking Bill.

Donald Kennedy, editor of *Science* (himself a one-time Democratic Carter Administration public servant), was active in the campaign against the Bush administration for being 'anti-science'. When asked what had led to this view among so many American scientists, Kennedy pointed to two issues: climate change and stem cells (Smith, 2005). This political campaigning might have played a role in his journal accepting an infamous paper by Hwang Woo Suk (Hwang *et al.*, 2005), later found to have faked his results, because the paper had considerable political significance. The Bush administration had in 2001 decided to limit federal funding for embryonic stem cell research, and the House of Representatives voted on 23 May 2005 to override this executive decision. This was helped by the timely publication in the on-line *Sciencexpress* a few days earlier of a landmark paper by Hwang *et al.*, reporting a very promising breakthrough in the efficiency of production of lines of human stem cells, suggesting that stem cell research held great promise. What is more, the paper was by a Korean team, which suggested that the US was being passed by for leadership in this area as a result of the executive decision. *Science* featured the paper (which was not published in the hard-copy journal until after the House vote) in a news story (Vogel, 2005), and the story featured prominently in the newspapers on 19–20 May – just before the vote.

Unfortunately, both this paper and another published by *Science* the previous year (Hwang *et al.*, 2004) were by Hwang, who within a year was dismissed from his post for falsifying this research and illegally obtaining human eggs for the experiments. As one correspondent to *Science* noted (Martin, 2006), 'If the *Science* editorial staff had paid more attention to the science and less to the sensation . . . the impact of this sorry affair might have been much less.'

The explanation of the reaction from the political Left in the US to Bush's 'sound science initiative', as we shall see, lies in the way in which values or culture intersects with the perception of risks that arise from environmental science, for it is in our cultural dispositions to risk that the partisan differences are based. As we saw with the reaction to Lomborg's book in the previous chapter, it was the framing of the science that most offended his critics, and (for example) his statement of the 'facts' of species extinction rates was even greater than that of one of his critics. This is not a trivial point for our argument here, because cognitive science suggests framing has important effects on human choice (De Martino *et al.*, 2006), and the influence of frames in producing bias is thought to be greater when information is either incomplete or overly complex, because we tend to rely more upon simplifying heuristics rather than extensive algorithmic processing under such conditions (Gilovich *et al.*, 2002), both of which are usually met with the 'virtual science' we are considering here. Many 'activist' environmental scientists (such as those we have studied here) seem largely unaware that it is their cultural views (or myths) of nature that drive their particular 'take' on science. And (as we shall see later in this chapter) it is the virtual nature of the science that makes such cultural dispositions relatively more significant in determining risk perceptions, but we can note here that this conclusion is supported by cognitive science.

Before we turn to such matters, however, we must first deal with some questions of the philosophy and sociology of science, in discussing by what criteria we might judge scientific information and various fallacious and other rhetorical devices that are often used to substitute for 'sound science'. In this chapter, I discuss some criteria by which we might judge the quality of scientific knowledge, and the ways in which virtuous causes can undermine such science. I also provide a brief case study of the 'nuclear winter' saga I introduced briefly in Chapter 1 to show that such 'corruption' has been occurring since long before the cases studied in the present book. I then look at the politicization of science in the USA, noting that (in addition to the more obvious industry funding of conservative think tanks and institutes) there are also substantial resources directed towards those on the Left of US politics supporting virtuous corruption and the spinning of such science for political effect. We will then consider in the final chapter just how political and cultural factors can affect the conduct of science, especially to show that no side of politics has a monopoly on environmental concern. Liberal-left politics in the United States, for example, might be more ready to embrace much of the environmental agenda, but there is no reason why other, diametrically opposed political ideologies cannot also embrace environmentalism on their own terms. But there is, as we shall see, a reason for the resonance between those on the modern Left and those who embrace what might be broadly termed the neo-liberal agenda, and this explains the differences we tend to observe in contemporary politics.

RECOGNIZING 'SOUND' SCIENCE

Historians and philosophers of science have long discussed the basis upon which we might judge the quality of science. In these discussions, prescriptive criteria by which we might judge some piece of scientific information to be better or worse are often mixed up with descriptive accounts of how science actually proceeds, or has proceeded historically.

Karl Popper (1963, 1968) stressed the importance of falsifiability as the fundamental principle of scientific method; that which could not be at least in principle be falsified was unscientific for Popper, and knowledge should never be regarded as proven, but always subject to falsification, and therefore held to be true only provisionally. Paul Feyerabend, in contrast, advanced an anarchistic view of the advancement of science, where the strict application of rules could impede the advancement of knowledge. Feyerabend (1975) held that new theories came to be accepted not because they accorded with some canons of scientific method, but because the supporters of the theory made use of all manner of tricks – both rational and rhetorical – to advance their cause. Feyerabend saw in this anarchism a necessary protection against the inevitable intrusion of ideology or theology into science, to the detriment of progress.

Interestingly, Feyerabend argued that, whereas it had begun as a liberating movement, science had become a repressing ideology, from which society should be protected. Rather than privileging scientific advice about which problems are worthy of solution, Feyerabend argued it should be judged alongside other 'ideologies', especially because, he argued, success by scientists had traditionally involved such non-scientific elements as inspiration from mythical or religious sources. He therefore argued for a separation between science and the state (just as it was accepted that religion and state should be separated) and for the democratic control of science.

Before Feyerabend, Thomas Kuhn (1962) had argued that there was little evidence that scientists employed a falsificationist methodology, instead working within a series of paradigms which were supported and defended against challenges from competing paradigms until each paradigm collapsed – or withstood the competition. Both Kuhn and Feyerabend present perhaps a more realistic descriptive account of the way in which scientists actually work (and the cases in this book provide further evidence), but their descriptive accounts provide very little guidance as to how society might judge competing scientific accounts. If Feyerabend is correct, and scientists use all manner of tricks to convince us that they possess the truth, one wonders how society is to exercise the democratic control over science that Feyerabend prescribes. In particular, as science is becoming less and less comprehensible to non-scientists, Feyerabend's philosophy of science appears ultimately to be a prescription for technocracy, for scientists being 'on top' rather than 'on tap'.

Popper's philosophy of science appears to offer a preferable prescription for how we might judge science, how we might winnow the useful scientific wheat from the chaff of metaphysical and ethical beliefs. It is probably impossible for science to be conducted without social, economic and political factors exerting some influence, almost gravitational, on its trajectory. But accepting as science only those propositions which are falsifiable, and placing more credence in those which have survived repeated attempts at falsification would appear to be a reasonable basis for a prescriptive standard against which to judge the attempts of scientists to mislead the public – to advance their careers, their reputations, their metaphysical beliefs, their ethical preferences, their political agendas, and so on.

Feyerabend might well be correct in asserting that scientists do not actually behave as honourably as Popper suggests, but we have every right to require that they *should* do so. To adopt Feyerabend's 'anything goes' view of science as a *prescription* is to abandon ourselves to the hegemony of scientists, and to abandon any hope of exerting accountability over them. This is not to say that falsifiability is the only test of 'sound' science. We can also be wary of numerous fallacious lines of argument and apply various other tests.

Michael Fumento (1993), for example, has listed ten common logical fallacies and rhetorical devices often committed or used by those attacking science, or wishing to undermine it (and many of these were evident in the attacks on McIntyre and McKitrick, or Lomborg, or Castles and Henderson):

1. *Post hoc ergo propter hoc* ('after this, therefore because of it') arguments fallaciously assign causality for events to events which precede them.
2. Circular reasoning. Observations are often a rich source of theoretical conjectures, but such conjectures cannot then be tested against the same evidence from which they are drawn. (Climate models are usually tested against the data they been constructed to explain, and the species–area equation does likewise.)
3. Straw man arguments involve the construction of an opponent's argument in a flimsy way, as a mere pastiche of the real argument, in order to take it apart for rhetorical effect. (McIntyre and McKitrick's 'audit' was repeatedly misrepresented as a 'reconstruction'.)
4. *Argumentum ad hominem* arguments involve attacks on individuals rather than their ideas. (McIntyre and McKitrick were dismissed as non-scientists with all kinds of motivations.)
5. Non sequiturs involve leaps of logic, of drawing conclusions from statements which do not follow logically.
6. *Argumentum ad populum* uses the rhetorical device that many people accept some statement, therefore it must be true. As Fumento (1993, p. 283) puts it, 'Democracy is a very nice thing, but it doesn't determine truth.' Or as as Galileo put it much earlier, 'In questions of science the authority of a thousand is not worth the humble reasoning of a single individual.' (The IPCC relies upon the authority of consensus among its participants.)
7. The genetic fallacy involves bypassing the argument, and instead going after its origin, usually assigning motives to it. Thus it may well

be that our suspicions might be aroused by the fact that a piece of scientific research was funded by some person or corporation with a vested interest, but that fact in itself makes it neither true nor false. It would make a world of difference, for example, whether the researcher was entirely free to research and to publish without fear or favour, whether they published in peer-reviewed journals, and so on. (Most climate change sceptics have this charge levelled against them, including that by Lord May as President of the Royal Society.)

8. Either-or thinking often sees the mischaracterization of issues as if they were black-and-white, and the protagonists as either totally right or totally wrong. The world is rarely so simple that someone who questions one piece of information also rejects all the relevant science. (Most sceptics accept there is some anthropogenic influence on climate, but contest the amount, but are characterized as being in total denial.)
9. Shifting the burden of proof. This often occurs with some interpretations of the Precautionary Principle, which (properly) aims to prevent opponents of policy action using any residual uncertainty to forestall action. There are often demands that, unless a chemical or practice can be proved to be safe, it should be prohibited. This is essentially an argument from ignorance, that – because we do not know everything – 'shouldn't we err on the side of caution?' This sounds seductively plausible, but the answer is sometimes a resounding 'No' – depending on the benefits and the risks. Such attempts to shift the burden of proof sometimes involve demands for what is sometimes called the 'witchcraft defence': prove that this presents no danger (or that you are not a witch). Logically, one can never prove a negative, and fortunately the dunking chair or burning at the stake are now employed only metaphorically (and survival does not provide proof of guilt).
10. Irrational appeals. Various kinds of irrational appeals (to common sense, to authority, to emotion) can be rhetorically seductive, but they have no bearing on the matter of truth.

Irving Langmuir (1985), a Nobel Laureate in Chemistry, once suggested a number of criteria to identify what he termed 'pathological science'. He suggested that society could ignore claims of hazard under certain conditions, which were effectively a set of prudential 'alerts' as to when 'science' was weak. The first of these was when the maximum effect was observed by a process of barely detectable intensity, and the effect was largely independent of the intensity of the supposed causal agent. Second were those situations where the effect was close to the limit of detectability. Third, claims of great measurement accuracy (or of profoundness) persist in the face of mounting evidence to the contrary. Fourth, the theories fail the 'Occam's Razor' test of being the simplest explanation of the available information. Finally, any criticisms are met by ad hoc explanations, and the proponents always have an answer. Henry Bauer has argued that mistakes are part of the conduct of science and that it is wrong to suggest such errors amount to scientific misconduct – or are even pathological (Bauer, 2002). He suggests that the concept of 'pathological science' lacks justification in the contemporary understanding of science studies, having been first termed by Langmuir in a seminar in 1953 and published in 1968 (and again in 1985). Bauer presents some cases (N-Rays, polywater, cold fusion) that suggest that such errors can happen, but he considers they are not necessarily 'pathological'. But Langmuir's checklist is still a useful way of judging the quality of science, regardless of whether that which it reveals as suspect is the result of fraud, pathology or simply the errors of well-meaning scientists.

Any official endorsement of what might turn out to be 'mistakes' goes against the anarchism Feyerabend advocates and Bauer seems to accept. Official endorsement is exacerbated by the attempts to suppress dissident views we have seen with Lomborg and the sceptical climate scientists. Ehrlich and his associates are now the scientific establishment: Raven was AAAS President at the time he was seeking to have Cambridge University Press effectively burn Lomborg's book; Holdren began his term in 2006; Kennedy is editor of *Science*; Lord May has been Chief Scientist in the UK and President of the Royal Society.

Fumento's ten points or Langmuir's five characteristics can serve as useful checklists when looking for poor arguments about science. But how do we know good science when we see it? Feyerabend might be content to let 'anything go' in science, but society must constantly decide whether to place its trust in a particular pieces of science, in decisions it makes in politics and the court system. We usually rely upon methodological tests, and these have been enunciated by several decisions by the US Supreme Court, in developing rules as to what will be admitted as scientific evidence. In the *Daubert* case (Daubert, 1993), these rules pose questions about whether a piece of evidence has been tested or whether it is at least falsifiable, whether it has been peer reviewed and published, what the risk of error is, whether it has been accepted in the relevant scientific community, whether the theory has been based upon facts or data of a kind reasonably relied upon by experts in the field, and so on. Some of these might be seen as fallacious (acceptance by a community of scholars is a version of *argumentum ad populum*), but no single criterion is necessary and none is conclusive. They are simply a test to help establish the reliability of scientific evidence.

The *Daubert* tests were later modified in the case of *Joiner*[1] and that of *Kumho Tire*.[2] In *Joiner*, the Supreme Court held that the evidence had to rise above subjective belief or unsupported speculation, and that the evidence offered had to be reliable. In *Kumho Tire*, the Court essentially ruled that the less scientific the evidence, the more important it was to ensure the quality of the expert testimony. Debate on these matters continues.

The *Daubert* case saw arguments aired that were fundamentally related to questions of the philosophy and sociology of science (Orofino, 1996). One argument stressed a view of science that was both Kuhnian and which saw science as socially and culturally constructed, and argued against the use of peer-review as some kind of 'litmus test'. (Such an approach would have favoured plaintiffs and trial lawyers (and precaution-based regulators) in making it easier for them to have 'science' accepted that might sway juries when married with non-scientific arguments and rhetoric.) Another, different argument invoked Popperian falsifiability and the need for peer review and publication as a necessary, though not sufficient, requirement for establishing what constituted 'good science'.

There is a strange alliance between those who argue for a more relativistic view of science (on the one hand) and trial lawyers and environmental campaigners (on the other), both of which groups want to be able to use affective arguments to help their causes overcome the limitations of the evidence. Melnick (2005) has argued that *Daubert* seeks to exclude 'valid science' from the courtroom, particularly that vital to the plaintiff's case in toxic tort legislation, because judges were unable to understand the methodology of contemporary science. But it is not just toxic torts, but also toxic regulation that some feel has been limited by a *Daubert*-style insistence on high scientific standards.

A revealing history of the relationship between science and regulatory policy in the USA is co-authored by Lynn Goldman, Clinton assistant administrator in charge of toxic substances in the EPA (Neff and Goldman, 2005). Neff and Goldman provide references to 'scientists' such as Devra Davis, who turns out to be attached the environmental group Natural Resources Defense Council (and was one of the posse attacking Lomborg for *Grist*), while frequently attempting to tar others for having accepted 'industry funds' – all pretty standard activist material, but surprising in a peer-reviewed journal. But the 'history' conveniently ignores all the factors that led to the move to 'sound science' gaining such traction: the awareness that most foods naturally contained carcinogens (which made an unworkable nonsense of the old Delaney Act prohibiting the sale of food containing carcinogens); the awareness that toxic substances regulation had proven cost-ineffective by chasing purity from lower-order risks (as found by Tengs, *et al.*, 1995); and perhaps most tellingly (since Neff and Goldman mention the Food Quality Protection Act 1996), the fraudulent science on endocrine disrupting chemicals that drove that legislation while Goldman was in charge of toxic substances regulation by the Clinton administration. And they fail to include any discussion of hormesis, which by the time they wrote their paper had made the error of linear dose-response models underlying carcinogenicity regulation all too apparent (Calabrese and Baldwin, 2003a).

Goldman was a signatory to the UCS statement against the Bush administration's 'sound science initiative', and from her article with Neff it is quite clear that she does not regard science as having been politicized under the Clinton administration. Yet a significant factor in science being politicized under the Bush administration was the politicization of science during the Clinton administration – captured in the title of an editorial by Donald Kennedy in *Science*: 'Well, they were doing it too' (Kennedy, 2003).

Interpretations of the list of criteria we might test 'science' against, vary from discipline to discipline. For example, in most disciplines, 'peer review' means double-blind refereeing, where the identities of the author(s) and referees are unknown to each other – to guard against *argumentum ad hominem*, or the genetic fallacy, or the seductiveness of authority. Others are less rigorous, and the reliability of findings published in such journals is regarded as less reliable, since these factors might come into play. As we saw, in many journals in geophysics (such as *Geophysical Research Letters*), for example, the identity of authors is customarily revealed to referees, apparently in the interests of facilitating debate among authors and reviewers. There is some merit in this, but we are not always talking here of arcane discussions which might occur among scholars simply seeking to understand the past. *GRL* has published some of the key papers on the science of climate change, and thus the participants are not simply in the business of interpreting the world, but of changing it, and some of them have well-developed policy preferences which give the findings of climate science a particular policy meaning (positive or negative). Such refereeing practices might have been appropriate when natural scientists were simply trying to get at the truth about the geological history of the earth, but they represent a problematic foundation upon which to base science which informs policy with such consequences as climate change.

Compare that with the practice in medical science with the protocols which must be followed for testing for drug efficacy. There, not only is the process of journal publication governed by double-blind conditions, but so is the conduct of research – and rightly so, since millions and sometimes billions of dollars are at stake in drug approvals processes. A double-blind study means there are four separate research teams, each being prohibited from having any contact with any other team. Ideally, they will be located at different universities, in different parts of the country. One team defines the study and makes up the medications, the real doses and the controls. The second team administers the drugs to the patients. The third team independently assesses the effect of the medications on each patient. The fourth team takes the data provided and performs the statistical analysis. This is a very costly process, but the US Food and Drug Administration must try to ensure the information upon which it bases registration decisions is as reliable as possible. There are similar standards agreed in the OECD for good laboratory practice (GLP) which provide for mutual acceptance of data (MAD) in chemical testing, to harmonize different national approaches and minimize cost and delay.

This strict regime is justified by the fact that the stakes are high, but it underscores the measures which are often taken to ensure that poor science does not lead to poor policy decisions. Adherence to such a strict regime does not guarantee that results will be correct, but we can have much greater faith in those results than with pieces of science where there is no separation between hypothesis generation, data selection and analysis – and where double-blind refereeing does not occur before publication.

One problem for traditional checks and balances in the process of science is that they assume a different world to that which now exists. The ultimate in peer review is double-blind peer review, where the identity of neither the author(s) nor the reviewers is known to each other. Even with double-blind anonymity, the editor has considerable power (and responsibility) in allocating papers to reviewers, and allocation to those whose work is either explicitly or implicitly criticized by a paper under review can mean a difficult passage through the editorial process, just as a reader whose work (and reputation) would be enhanced by positive citation might be tempted to be more sympathetic.

There is a dilemma here: those who have published in a field are best suited to pass judgement on the quality of a paper, but they have a clear conflict of interest when reputational issues are apparent and success in grant applications, appointments and promotions depend upon professional standing, measured not just by publications but also by measures such as citations. There is a strong ethical imperative for scholars to rise above such conflicts, but there is a strong temptation for them to do otherwise. Editors and reviewers can thus act as 'gatekeepers' in specific areas of knowledge, deciding who is 'qualified' to enter. Such areas of scientific endeavour can, for the insiders, become 'club goods', access to which is controlled and the benefits accruing to which must be defended from challenges which might devalue their worth.

Ironically, a case can be made for the identity of reviewer to be known to the author under conditions where the reverse is true, as this would act as a discipline upon reviewers who might otherwise act improperly in exercising their judgement. The current system protects only editors and reviewers, though the counter argument is that editors might find it more difficult to recruit reviewers to the unpaid task of refereeing. But even double-blind refereeing is no longer a guarantee of impartiality – if, indeed, it ever was – because the nature of science in the contemporary world has substantially undermined its effectiveness. There have been two reasons for this: the increasing specialization of knowledge; and the communications revolution that lies at the heart of globalization.

Once, it was possible for individual scholars to sustain expertise across most of their discipline, and sometimes beyond, into cognate disciplines. Today, there is such a degree of specialization that there are substantial areas of knowledge where the qualifications to participate are not shared even across disciplines: the analytical skills required are increasingly specialized and the literature has likewise expanded so that most scholars work in much smaller research areas, lacking either the skills or detailed knowledge of the relevant literature (ignorance of which will likely produce a report recommending rejection). An expert can be defined as someone who choses to remain ignorant about many things in order to know everything about one area. The size of the sacrifice which is now the necessary price of expertise continues to grow as knowledge grows.

So, as we noted, if one looks at the scientific literature on a scientific area such as the use of proxies for reconstructing past climate histories back before an instrumental record is available – the technique at the heart of the 'Hockey Stick' controversy – one finds a recurring list of names: K.R. Briffa, E.R. Cook, T.J. Crowley, J. Esper, P.D. Jones, M.E. Mann, A. Moberg and T.J. Osborn dominate the literature, supplemented by a few others. Any editor looking for reviewers for a paper on this subject would likely (quite justifiably) turn to two of those named on this list. But the small size of this list means that there is not a wide range to choose from.

This difficulty is exacerbated by globalization. In the past, communication was both more difficult and substantially more expensive than it is now. There was probably a reasonable chance that there was a degree of independence among the leading scholars in any field, because attendance at key conferences was so expensive as to limit contact, and 20 years ago exchange of drafts of papers had to occur via, at best air mail and then later facsimile. Data could only be made available to other researchers with great difficulty, on cumbersome computer tapes and then disks. Now, papers and data can be exchanged around the globe at close to zero cost and in the twinkling of an eye. Pre-publication copies of papers appear on websites. Ideas spread through Internet chat groups. Air travel costs a fraction of what it once did before jumbo jets and fuel-efficient engines. This has transformed radically the nature of scholarship: just as the Internet allows 5000 believers with UFO conspiracy theories to find each other, so too can the small groups of scientists in any one area meet each other, share ideas and collaborate.

So we find that the leading scholars in climatic reconstructions using dendrochronology are not independent of each other (as they might have been 30 years ago); they are not only known to each other, but they have collaborated on published research, as Wegman found. Using the simple technique employed for many years by Marxist scholars to demonstrate the existence of a 'ruling class' by means of mapping out interlocking directorships of corporate business, we can show just how close is this circle of climatic dendrochronology. The Marxist analysis always had its problems: directors might sit on the same board, but have differences with others who share directorships, for example, so coherence could not be merely assumed. But here, we are talking about public evidence of collaboration evidence by multiple joint authorship.

As we saw in Chapter 3, almost the entire Hockey Team was assembled to attempt to rebut the 'heresy' of Soon and Baliunas in 2003 (Mann *et al.*, 2003), and the extent to which modern communications technology makes possible collaboration over long distances can been seen when the institutional affiliations of the authors are listed. Michael Mann was at the University of Virginia, in Charlottesville, USA; Caspar Amman and Kevin Trenberth were at the National Center for Atmospheric Research, Boulder, Colorado, USA; Ray Bradley the University of Massachusetts, Amherst, USA; Keith Briffa, Philip Jones and Tim Osborn, the University of East Anglia, Norwich, UK; Tom Crowley, Duke University, Durham, North Carolina, USA; Tom Wigley, the University Corporation for Atmospheric Research and NCAR, Boulder, Colorado, USA; Malcolm Hughes and Jonathan Overpeck, the University of Arizona, Tucson, Arizona, USA; Michael Oppenheimer, Princeton University, Princeton, New Jersey, USA; Scott Rutherford, the University of Rhode Island.

Ironically, the formation and operation of the Intergovernmental Panel on Climate Change has facilitated interaction among such scientists, both through electronic communication in authorship and the review process for the IPCC and by bringing scientists into more frequent face-to-face contact. And, of course, the provision of large research budgets has meant that many work full-time in dedicated research institutions (the funding of which depends on the continuation of concern) and are more able to attend conferences.

Cheap, fast communication permits the scientific equivalent of a phenomenon noted by criminologists with users of mobile phone technology (especially text messaging): 'swarming'. Swarming has seen large groups of youth descend on one place, with riotous intent, as well as more benign social gatherings where being at the designated location at the appointed time is the 'thrill'. The response to Lomborg's heresy was one example of such 'swarming', with Ehrlich's associates swarming like white blood cells to the site of the infection. Yet the same revolution in communications and information technology which makes it possible for defenders of a scientific orthodoxy to muster across considerable space has also transformed the ease with which data can be archived and made available to other researchers – and has facilitated the ease with which such researchers can subject it to audit and verification. The Hockey Stick case shows how the near-zero cost of information exchange can exert a discipline over the conduct of science – as long as adequate disclosure and transparency are the norms. The Hockey Team's refusal to disclose data and methods, and Michael Mann's deletion of files and barring of McIntyre's ISP address look exceptionally churlish. Not surprisingly, many journals are now beginning to require data supporting scientific publications to be archived, though the practice is not common in climate science.

This increased ease of communication undoubtedly has advantages in terms of collaboration and transparency, but it also contains some dangers that have not received much attention. One of these is the possibility of a virtual variant of what Irving Janis (1982) once called 'Groupthink' – the construction and maintenance of a particular view (quite possibly wrong) and its defence in the face of external challenge. Groupthink is a kind of collective 'cognitive dissonance' (as Leon Festinger (1962) described it) where inconvenient information is rationalized away. Groupthink is a danger because the group can operate virtually, with connections largely unseen, and can establish a monopoly or at least epistemic hegemony over particular areas of knowledge. Thus, when confronted with challenging research, the wagons are not just circled, but circled around the heretics, so that fire can be concentrated on them.

The operation of any group necessarily involves politics. Be it a relatively small group (such as the Hockey Team) or a large formal organization such as the IPCC, decisions must be made to resolve differences of opinion, and these processes are the very stuff of politics. They might be made according to agreed rules (as with the IPCC) or by less formal means, but even in the application of formal rules basic issues of values on such matters as justice and the distribution of power will inescapably be part of decision-making. The use of majority voting or the development of a consensus are inherently political processes, so consensus science is inescapably political science. We must therefore consider the social and political context within which science is conducted.

That context appears to have a stronger grip on science in the United States than elsewhere, and this appears to be the result (at least in part) of earlier concerns over the prospect of nuclear war, so that (while the risks of nuclear annihilation appear to have abated with the end of the Cold War) the politicization of science persists. We can see this by examining the case of 'nuclear winter', the formation of the Union of Concerned Scientists and the activities of numerous charitable trusts there which constitute a relatively unseen part of the value-slope that is applied to science, especially environmental science in the United States. Such foundations provide substantial funds that are at least the equivalent of those supplied by business and conservative foundations, and consideration of their activities suggests that if there is a 'war on science' being waged by the Bush Administration, it is by no means a one-sided affair, and hostilities might well have been commenced (or, at least, escalated) by the other side.

THE NUCLEAR WINTER SAGA

The nuclear winter saga saw the participation of many of those involved later in climate science and the attack on Lomborg, either alone or in collaboration. Paul Ehrlich collaborated with Peter Raven, the AAAS President who sought to prevail on Cambridge University Press to withdraw Lomborg's book, on the biological consequences of nuclear winter (Ehrlich *et al.*, 1983). Ehrlich and Donald Kennedy (whom Ehrlich later introduced to readers as the new editor of *Science*) collaborated on a book on nuclear winter (Ehrlich *et al.*, 1984). Stephen Schneider, having emerged with concerns over cooling or warming in the 1970s, joined the nuclear winter issue, though his modelling helped dampen it down (Thompson *et al.*, 1984; Covey *et al.*, 1984; Schneider *et al.*, 1986). Moreover, not only were many of those concerned with climate change represented in nuclear winter science, but so too were many of their critics, with later climate change sceptics such as Garth Paltridge (Barton and Paltridge, 1984), Sherwood Idso (1984, 1986) and Russell Seitz (1985), for example, all adding sceptical voices to the literature.

Brian Martin (1988) suggested that the nuclear winter alarm reflected the resurgence of the peace movement in the early 1980s, with peace activism spreading throughout many organizations and occupational groups, including doctors, scientists and engineers. The science began with a deliberate decision by the Swedish Academy of Sciences to address the nuclear issue, with the editors of the Academy's journal *Ambio* deciding to publish a special issue in 1982 to cover the effects of nuclear war. Paul Crutzen, later to win a Nobel Prize, was asked to write on the effects of nuclear war on the atmosphere (Crutzen and Birks, 1982).

This scientific concern came after an exercise in popular journalism. Jonathan Schell wrote a series of articles in the *New Yorker* magazine in 1981 claiming that nuclear war could cause extinction of human life, principally through destruction of stratospheric ozone. Schell's articles were inspired by the peace movement and taken up enthusiastically by it, and were then developed into a popular book – *The Fate of the Earth* (Schell, 1982). But Schell's argument about ozone depletion had already been rendered redundant by developments in nuclear strategy, which made it highly unlikely that a nuclear war would deplete stratospheric ozone.

Crutzen was an expert on the effect of nitrogen oxides in regulating the amount of ozone in the stratosphere, and he and Birks used computer models dealing with stratospheric ozone to determine the effects of a nuclear war. The large multi-megatonne warheads of the 1950s were being replaced by larger numbers of smaller warheads, however, and fewer nitrogen oxides would transported up into the stratosphere. The Crutzen and Birks' modelling therefore did not predict a significant reduction in stratospheric ozone using the reference scenario *Ambio* had provided. Rather, they focused on the smoke released by fires caused by nuclear attacks and performed calculations which showed that smoke could absorb much of the incoming solar radiation, resulting in a 'twilight at noon'.

This idea was quickly developed into 'nuclear winter' by the authors often referred to as 'TTAPS' – Richard Turco, Owen Toon, Thomas Ackerman, James Pollack and Carl Sagan (Turco *et al.*, 1983) – who calculated dust blocking the sun would lead to massive cooling of the earth, and Ehrlich *et al.* (1983), who estimated the catastrophic biological effects of all this in a paper which followed immediately in *Science*. This was conjecture masquerading as science in the service of the noble cause of nuclear disarmament, as Crutzen was later to make perfectly clear in his Nobel Lecture, when he stated:

> Although I do not count the 'nuclear winter' idea among my greatest scientific achievements (in fact, the hypothesis can not be tested without performing the 'experiment', which it wants to prevent), I am convinced that, from a political point of view, it is by far the most important, because it magnifies and highlights the dangers of a nuclear war and convinces me that in the long run mankind can only escape such horrific consequences if nuclear weapons are totally abolished by international agreement. (Crutzen, 1995)

The papers by TTAPS and Ehrlich *et al.*, continued this concern, and amplified the risk by making numerous assumptions which embodied the worst case for the possible effects of a nuclear war. Martin pointed out that they did so by such devices as assuming targeting scenarios that would generate enough dust and smoke to produce a nuclear winter – reminiscent of the SRES scenarios in climate change later. They also suggested there was a sharp threshold above which severe nuclear winter effects would be triggered; there was little scientific justification for this assumption, but it was convenient for policy purposes, especially as Sagan had suggested that nuclear arsenals should be reduced below such a threshold.

TTAPS also used a one-dimensional model which showed dramatic temperature reductions over land but little change over oceans. The authors did discuss the moderating effect of the oceans in the text, but most readers and commentators concentrated on the tables and the abstract, where the extreme land results were highlighted. Ehrlich *et al.*, then focused on the land results from TTAPS over the whole globe in assessing the biological effects of nuclear winter, and suggested all manner of disasters from nuclear war, including, for example, decreases in stratospheric ozone and resulting increases in ultraviolet radiation – once the smoke and dust had cleared, of course – while ignoring the point above (acknowledged by Crutzen and Birks) that changes in the size of warheads had largely removed this threat. They even raised (only in the summary and conclusion, rather than in the body of the text) the possibility of human extinction, without explaining precisely how the whole of humanity might die, and failed to mention factors which might ameliorate problems. Martin concluded that the TTAPS and Ehrlich *et al.*, papers are not 'value-neutral' pieces of research, but 'pushed' certain conclusions on readers through technical assumptions in model construction, selection of evidence and highlighting of results.

This 'science' supported the previously expressed political preferences of many of the participants. For example, as noted above, Carl Sagan had previously argued for 'deep cuts' in nuclear arsenals to reduce them below the nuclear winter threshold, and Barrie Pittock, a prominent promoter of alarm over nuclear winter in Australia (and later promoter of global warming) had argued against Australia's nuclear alliance with the United States. Martin also suggested that some Soviet nuclear winter scientists close to Gorbachev used nuclear winter arguments to influence Soviet disarmament proposals, emphasizing the worst effects of nuclear war even more than Western scientists. In a close parallel with global warming later, Martin also noted that the proponents held the scientific high ground, carrying the weight of numerous eminent recommendations, prestigious journal publications and scientific committee endorsements and able to portray their results as purely scientific and above politics. Their critics, as with global warming sceptics, were not in a position to present alternative model results, but could only make methodological criticisms and point to uncertainties. While they largely argued within a scientific context, they were in a much weaker position, and more often raised overtly political criticisms.

The issue also saw an attack on one critic, Russell Seitz, that closely paralleled that of Mann *et al.*, on McIntyre and McKitrick over their criticism of the Hockey Stick, and those on Lomborg. In response to criticisms from Seitz, the TTAPS authors defended their work by referring to other studies which had confirmed their original claims (none, of course, based on observations), but also launched an attack on Seitz himself, challenging his standing as a scientist. Turco claimed he was a stock investment consultant dabbling in atmospheric physics, who was not the principal author of a single peer-reviewed scientific work in any technical field. TTAPS contrasted this with the credentials of nuclear winter scientists to whom (together with Crutzen and Birks) the American Physical Society had awarded a prize for their research on the nuclear winter theory. This attack on Seitz was not only *ad hominem* but also inaccurate: Seitz at the time held an appointment at Harvard and had worked at R.J. Edwards, not as a stock consultant, but as Director of Technology Assessment. He was also principal author of peer-reviewed scientific publications.

One issue in the dispute between TTAPS and Seitz described by Martin is particularly illuminating. Seitz claimed in an article published in the *Wall Street Journal* that Kosta Tsipis of the Massachusetts Institute of Technology quoted a Soviet scientist as saying 'You guys are fools. You can't use mathematical models like these to model perturbed states of the atmosphere. You're playing with toys.' TTAPS in a November 1986 letter to the *Wall Street Journal* stated that:

> A negative comment on mathematical modeling allegedly uttered by a 'Soviet scientist' (indisputably V.V. Aleksandrov of the Moscow-based Climate Modeling Center, the only Soviet at the April 1983 Cambridge review meeting referred to by Seitz), and prominently displayed in a box by the *WSJ*, was never made. The transcript of the meeting shows no such remark, and Kosta Tsipis of MIT, whom Seitz claims as his source, flatly denies the whole thing. (Martin, 1988, p. 330)

Tsipis, in a memo of 5 January 1987, entitled 'Regarding: Seitz vs. Sagan', gave his account:

> When Russell Seitz came to talk to me about Nuclear Winter, I recalled that in the AAAS Meeting (in Cambridge Mass.), a Russian scientist got up and said that we cannot use climate models as if the nuclear war itself would not disturb the atmosphere. The discussion at that point had evolved around the 1-D [one-dimensional] model. Mr. Seitz mentioned this in his Wall Street Journal article, but in a context that implied that the Soviet scientist was referring to all 3-D

> models, quite generally. Subsequently, I had a telephone call from Carl Sagan who wanted to know what I had said to Seitz. During our conversation, two things became clear: a) that Seitz had confused my statement to mean that it referred to a 3-D model; b) that it would be very difficult to explain to the readers of the W.S.J. the distinction. For this latter reason, we agreed that Carl should simplify his response by saying that I deny discussing the 3-D model with Seitz. In Carl's letter-response in the W.S.J., this statement was further simplified. (Martin, 1988, p. 330)

It was interesting that Sagan and Tsipis agreed upon an incorrect account to avoid confusing readers, but Seitz later wrote to Martin stating that Tsipis' original remarks had been recorded, that the clear context was 1-D models, and that he was not aware of any confusion between 1-D and 3-D models in the text of his *Wall Street Journal* article.

Nuclear winter, of course, faded from concern with the end of the Cold War, though (as we have seen) it depended on old targeting scenarios that were no longer current, kept alive in the virtuous cause of ending the nuclear arms race. Ultimately, that result came about not by the international agreement activist scientists such as Crutzen advocated, but by the collapse of the Soviet Union, at least partly the result of its inability to afford to remain in the arms race, not because they sought not to compete. What the nuclear winter episode indicates is that nuclear concerns had politicized science, and because such concerns had become institutionalized in organizations such as the Union of Concerned Scientists, the politicization of science outlived the Cold War.

UCS began in December 1968 at MIT with a Founding Document in the form of a Faculty Statement in response to the Vietnam War signed by 50 senior faculty members, including the heads of the biology, chemistry and physics departments, and later circulated to the entire faculty for endorsement. This was followed by the founding of the Union of Concerned Scientists in early 1969. The Founding Document stated that:

> Misuse of scientific and technical knowledge presents a major threat to the existence of mankind. Through its actions in Vietnam our government has shaken our confidence in its ability to make wise and humane decisions. There is also disquieting evidence of an intention to enlarge further our immense destructive capability. (UCS, 2006)

A prominent aim (the second listed) of UCS was 'To devise means for turning research applications away from the present emphasis on military technology toward the solution of pressing environmental and social problems.'

UCS has played an active role in anti-nuclear and environmental politics, issuing a 'World Scientists' Warning to Humanity' in November 1992 and a 'World Scientists' Call for Action' at the Conference of the Parties to the FCCC at Kyoto in December 1997, at which the Kyoto Protocol was agreed. It also took the lead in opposing the Bush Administration's 'Sound Science Initiative' issuing a statement on 18 February, 2004, signed by over 60 prominent scientists, voicing their concern over the 'misuse of science' by the Bush administration. Among the signatories were several names active on the issues of conservation biology and climate change we have studied here, including Paul Ehrlich, John P. Holdren, Stuart Pimm and Kevin Trenberth, a leading climate scientist at the National Center for Atmospheric Research. Other scientists took part in UCS's role in attacking Bjorn Lomborg and Stephen Schneider, for example, has represented it at climate negotiations.

The Union of Concerned Scientists goes beyond issuing warnings to humanity and calls to action. For example, it also cooperated with Public Employees for Environmental Responsibility (PEER) in distributing a 42-question survey to more than 1400 USFWS biologists, ecologists, botanists and other science professionals working in Ecological Services field offices across the United States seeking their perceptions of scientific integrity within the USFWS, as well as political interference, resources and morale. This points to a politicization of the administrative branch of government in the United States which predated the presidency of George W. Bush, and which provides important context to the claims that Bush has waged a 'war on science'. The argument here is *not* that Bush has not politicized science; he and his administration are clearly hostile, for example, to use of human cloning for stem cells. Rather, the point is that science in the US had already been politicized, and this facilitated the further politicization of science by the Bush administration, because it could see that previous administrations had allowed politicization in line with their ideology, and the Bush administration thought it acceptable for them to do the same.

POLITICAL SCIENCE IN THE US GOVERNMENT

The executive in the United States is much more heavily politicized than in parliamentary systems of government, with the top levels of the public service changing with each change of president. Just how far down partisan appointments reach is variable, but it has probably increased with time (and was extended deeper with the Regan Administration). What is notable is that the senior political appointments to environmental agencies during the Clinton Administration included large numbers with past associations with environmental groups. Secretary of the Interior Bruce Babbit, for example, brought with him an association with the League of Conservation

Voters and EPA Administrator Carol Browner one with Citizen Action. Many other senior staff had past associations with groups such as the Wilderness Society (Director of the Bureau of Land Management, Jim Baca and Director of the Office of Management and Budget, Alice Rivlin). Staff with associations with WWF, Natural Resources Defense Council, Friends of the Earth, the Nature Conservancy, Sierra Club, and Audubon Society held positions in agencies such as OMB, Department of Agriculture, Interior, Bureau of Land Management and EPA. Tim Wirth, former Democratic Party Senator from Colorado, who had allowed his 1988 committee hearings on climate change to be orchestrated by James Hansen and Friends of the Earth for maximum political impact (see p. 173), was given a prominent position in the State Department and exerted considerable influence on both the negotiation of climate treaties and the IPCC.

These appointments came on top of a substantial politicization of lower levels of the public service, with environmental groups formed specifically to change government policy from within (regardless of the party in office). The Forest Service Employees for Environmental Ethics was formed in 1989 and had around 12 000 members in 1998 and a budget of $900 000, according to the *Encyclopedia of Associations* (Gale Research, 1998). It works to advance conservation and oppose 'overuse' of public land by timber companies, mining firms and cattle owners. Its Executive Director Andy Stahl was involved in sponsoring the production of peer-reviewed science to support the Spotted Owl campaign to reduce old-growth logging in the Pacific Northwest, and was also the person claiming that the planting of Canadian lynx fur was a test for laboratories rather than a fraudulent attempt to secure further protection under the Endangered Species Act.

Nature editorialized in support of those who had faked evidence of Canadian lynx (*Nature*, 2002; Dalton, 2002), and was quickly taken to task for supporting this unjustified planting of samples (Mills, 2002) by a researcher involved in this project and another, the integrity of which was impugned by the fakery (Schwartz *et al.*, 2002). The Forest Service had in 1998 contracted John Weaver who worked for the environmental group, the Wildlife Conservation Society. He had reported lynx hair in both Oregon and Washington in areas where nobody expected them. These results were used in a Forest Service application for listing the lynx as an endangered species, but the samples were later found to be from bobcats and coyotes. Further evidence could not be found. Then, in the 1999 and 2000 survey seasons, seven employees sent in samples labelled as wild lynx. While they claimed to be testing the lab, they were discovered only because one employee blew the whistle the day before he retired (Strassel, 2002).

A similar organization, for federal employees in other agencies, Public Employees for Environmental Responsibility, was founded in 1993, and operates with a budget a little smaller than FSEEE, to work with and on behalf of employees to effect change in the way resource agencies conduct environmental management. PEER encourages whistle blowing and 'anonymous activism', and distributes 'Undercover Activist' boxer shorts. When the efforts of PEER and FSEEE operating at lower levels in federal organizations are added to the capture of the upper levels by executive staff linked to activist groups, and the spread of activist conservation science in the Society for Conservation Biology are considered (see below), the extent of the politicization of environmental science and its harnessing to politics in the United States can be appreciated.

The point here is not that this is particularly unusual, but to place in context the claims which greeted the overturning of this established order by the Bush Administration when it came to power in 2001. The move to require 'sound science' was attacked by groups such as the Union of Concerned Scientists as representing the politicization of science. But the fact of the matter was that environmental science in the United States federal government was already highly politicized and the scientists whose views were informing policy making were very much activist scientists. The Bush Administration was merely sweeping aside the status quo and turning the tide in a different direction. And it might well have been true that some industries welcomed this change and donated to the Bush campaign in anticipation of it, but this ignores the extent to which the *status quo ante* under the Clinton Administration was also heavily supported financially, though proportionately more by liberal-inclined charitable trusts.

The point of this is that science is the poorer for its politicization by both sides of politics – both for its own sake, and for what this means for public policy. While the thesis of this book is that noble cause corruption gives as much cause for concern about the reliability of science as the pernicious influence of money and interests such as reputation (and that such factors might be more pernicious because they are less obvious), many noble causes in the United States are themselves heavily subscribed financially. There is substantial financial backing for causes such as environmental protection, not through the small donations of millions of members and supporters, but through the grants of numerous charitable trusts – ironically, established on the basis of wealth generated by industries such as oil.

This picture of substantial funding using loopholes to allow tax deductibility for political activity and for 'black marketing' is all presented by numerous right-wing organizations which have assembled the funding trails through examining their IRS Forms 990 (available at GuideStar.org). Such groups are concerned to expose this unseen and 'undue influence' on the Left. Our concern here is not with the legitimacy of such influence activity, but simply to establish that it exists and is significant on the Left of politics, and among those supporting environmentalism, just as it exists on the Right and among those opposing environmentalism. The problem with most analyses of the way in which science is 'spun' is that they focus only on the more obvious attempts by industry to counter the claims of environmentalists, and ignore totally the extent to which this is also rife on the other side of politics.

Grant-giving organizations such as the Pew Charitable Trusts, W. Alton Jones Foundation and Surdna Foundation have ensured that environment groups in the USA are largely freed from the need to spend effort mobilizing resources. They make donations to environment groups, for scientific research into issues they wish to highlight, and also support organizations able to create and expand political support for change. One notable example of this was the campaign supported by the W. Alton Jones Foundation (now the Blue Moon Fund) to regulate endocrine-disrupting chemicals (EDCs) in the 1990s.

Alton Jones funded research at Tulane University that suggested that minute traces of EDCs acting synergistically could have a greatly amplified effect. This coincided with a book, ominously titled *Our Stolen Future*, which was co-authored by the director of Alton Jones (John Myers), a WWF chemist (Theo Colborn) and a journalist (Diane Dumanoski) and a widespread campaign for action. (Ironically, the early modern environmental movement had focused on overpopulation; now the threat was humanity's inability to reproduce!) The US Congress responded with legislation in the form of the Food Quality Protection Act 1996 – only for the Tulane paper subsequently to be formally withdrawn when it could not be replicated. The timing of the book, the paper and the campaign were seemingly no coincidence. Alton Jones also funded numerous environment groups to support the campaign for action.

Troy Seidle (2004) provides a scathing analysis of *Our Stolen Future*, the campaign over EDCs, and the passage of the Food Quality Protection Act. (Seidle, of People for the Ethical Treatment of Animals, was concerned about the impact of the demands of the legislation on the incidence of animal testing.) *Our Stolen Future* was hailed as the next *Silent Spring*, but it received critical reviews in both *Scientific American* (Kamrin, 1996) and *Science* (Hirshfeld, 1996). The Tulane study (Arnold *et al.*, 1996) appeared just as the media hype over *Our Stolen Future* began to fade, and it was endorsed strongly by EPA Administrator Carol Browner and EPA Assistant Administrator of Prevention, Pesticides and Toxic Substances, Lynn Goldman (who, ironically, was later to sign the UCS petition against the sound science initiative). Yet the paper quickly came under critical scrutiny because it could not be replicated (Kaiser, 1997; Ashby *et al.*, 1997), and it was formally withdrawn in 1997. The US Office for Research Integrity found subsequently in 2001 that there was no original data or other corroborating evidence to support the research results and conclusions reported in *Science*, and the authors were banned from receiving federal grants for five years.

The highly emotive politics that accompanies science on toxics has resulted in a substantial distortion of policy priorities. For example, Tengs *et al.* (1995) concluded from a study of 500 life-saving interventions that the median cost for a life-year saved was $42 000, and the median medical intervention cost $19 000 per life-year. But the median cost of toxics regulation was $2.8 million per life-year. It now seems that the regulation of chemicals for carcinogenicity has been based upon poor science. There is widespread evidence of hormesis – the theory that small doses are *beneficial* in that they induce *fewer* tumours, so that the dose–response curve is J-shaped (Calabrese and Baldwin, 2003a, 2003b). Whereas most modelling for chemicals regulation uses a threshold dose–response model, carcinogenic regulation has proceeded on the basis of a linear dose–response curve and has thus resulted in an expensive regulatory quest for levels of purity that are not only unattainable, but are probably harmful if successful.

The opportunity cost of such politicized science is that resources are misallocated and lives that could have been saved are lost, and such examples are not confined to biodiversity, climate change and toxics regulation. In April 2002, for example, *Nature* formally withdrew a paper by David Quist and David Chapela that it had published in November 2001. In it, they reported that modified genes in maize in the USA had crossed into Mexico and contaminated wild maize plants. Dr Quist had stated that this research showed that the benefits of GM crops did not outweigh the 'enormous' risks to food security (*Telegraph*, 10 August 2002). (The risk assessment was not, of course, Dr Quist's to make – nor was it within his field of expertise.) Publication of the paper had sparked protests over the methodology by 100 leading biological scientists (evidence, said activists, of a biotech industry vendetta), and it was later disowned by the Mexican government after its scientists were unable to replicate the results. It was something of a surprise, therefore, that Jorge Soberon, head of the Mexican delegation to the conference of the parties to the Convention on Biodiversity (CBD COP) told a meeting in The Hague in April 2002 that further tests by scientists from the National Autonomous University of Mexico and the Environment Ministry had confirmed the finding.

This no doubt played well at the CBD COP, as Mr Soberon stated: 'This is the world's worst case of contamination by genetically modified material because it happened in the place of origin of a major crop. It is confirmed.

There is no doubt about it.' Unfortunately for Mr Soberon, doubts persisted – at least until a further study reported in the *Proceedings of the National Academy of Science* could find no evidence of contamination (Ortiz-Garcia *et al.*, 2005). Perhaps not surprisingly, this research was reported by *Planet Ark* under the headline 'Gene-Modified Corn Gone from Mexico, Study Finds' (*Planet Ark*, 9 August 2005), suggesting that the contamination had been there, but had now mysteriously vanished, rather than the more obvious conclusion that the flawed Quist and Chapela paper had been wrong in finding that it was present.

Daniel Sarewitz (2004, p. 386) takes a constructivist position on science, arguing that advocates on either side of an issue are likely to exploit any uncertainty or competing scientific results to support their position, arguing (after Jasanoff, 1987, 1990, 1996 and Wynne, 1989) that science is inevitably embedded in a political context, and the boundaries between science and policy or politics 'are constantly being renegotiated as part of the political process' (see also Jasanoff and Wynne, 1998). Be that as it may, we need not accept his conclusion that all science is equally indeterminate, and equally subject to construction. Indeed, the very example Sarewitz provides has (since his piece was published) been settled because the contentious results could not be replicated – a somewhat 'realist' test of their veracity. (Critical realism, which holds that social construction is largely confined to the derivation of our knowledge, does not consign us to relativism.)

Sarewitz (2004, pp. 390–3) used the example of the research of Chapela and Quist that suggested wild Mexican maize had been contaminated by GM pollen to suggest that the controversy reflected the embedding of different values in genetic science. The practitioners of genetic science pointed to methodological flaws in Chapela and Quist. Environmental scientists were prepared to acknowledge the methodological flaws, but recognized 'that parts of the research had important implications for ecosystem behaviour, and as well that the research reflected such scientific virtues as replicability of results and the clever identification of a control case.' Ultimately, of course, the fact that the research could not be replicated settled the issue rather definitively, but the case is a salutary one, because Chapela and Quist's findings (unsurprisingly) reinforced the open hostility to GM technology, and the environmental scientists – in reading 'implications' from the 'good parts' of a flawed study – demonstrated a common problem for policy based on such science: the science itself frequently embodies the precautionary principle, so that double counting of precaution results when it is invoked in the policy process.

If we accept that science is largely self-correcting (as it was with these EDC and GM contamination cases), there is not a great cause for concern, but the provision of substantial resources to push the politics can lead (as in the EDC case) to premature regulatory action aided by the invocation of the precautionary principle. It is for this reason that the funding behind those scientists pushing noble causes *is* of concern, and I now turn to examine this phenomenon in some detail.

FUNDING NOBLE CAUSES

Alton Jones, active in the endocrine disrupting chemical episode, was established in 1944 by W. Alton Jones of the Cities Service Company (owner of utilities and oil pipelines) with the purpose of financing artistic and cultural activities. It became increasingly political in the 1980s over nuclear issues, and then began increasingly to fund environmental causes in the 1990s with the appointment as director of zoologist Myers, who had previously worked for the Audubon Society. In 2001, Alton Jones was restructured into three separate foundations with two (the Oak Hill Fund and the Edgerton Fund) taking somewhat more practically-applied directions, with the direction under Myers being continued by the Blue Moon Fund under Jones' grand-daughter Diane Edgerton Miller (president and CEO) and daughter Patricia Jones Edgerton (treasurer). Blue Moon funds most of the environment groups: Sierra Club, Natural Resources Defense Council (NRDC), Environmental Working Group (EWG), Greenpeace, the Nature Conservancy, Friends of the Earth (FOE), Worldwatch Institute, Environmental Defense Fund (EDF), WWF, League of Conservation Voters, Association of Forest Service Employees for Environmental Ethics, and World Resources Institute (WRI).

The Pew Charitable Trusts consists of seven separate funds established between 1948 and 1979 by the four children of Joseph N. Pew, the founder of the Sun Oil Company. It has assets in excess of $4 billion and distributes over $250 million in grants each year, with the Environmental and Health and Human Services Programs being the largest. Pew was a conservative, but his wealth has been directed to progressive political causes, beginning in the late 1980s under the administration of Thomas W. Langfitt. Pew has held investments in oil companies, including Chevron, Atlantic Richfield, Phillips Petroleum and ExxonMobil, yet supports campaigns against ExxonMobil and other oil companies, and supports the main environment groups: FOE, Sierra Club, WRI, WWF, Wilderness Society, WWF, EDF, NRDC, the Nature Conservancy, EWG, and Worldwatch Institute.

The Pew Charitable Trusts spent $4.5 million successfully lobbying the Clinton White House to close roads on federal lands, which it did by executive order shortly before leaving office. Pew defended lobbying for policy change by citing free speech, and escaped with its 501(c)(3) tax exempt status intact because it is only prohibited from lobbying for legislation, not executive orders. The House Subcommittee on Forest Health investigated the decision and found the decision was made improperly, in apparent violation of the due process rights of affected parties as well as the statutes enacted to protect those rights (Administrative Procedures Act and Federal Advisory Committee Act). The Committee reported that this 'structured relationship' between the Administration and environmentalists was of serious concern, but it considered that more significant was the lack of any evidence of even a token effort by the Administration to involve other interested parties.

The Surdna Foundation was established in 1917 by John Emory Andrus of the Arlington Chemical Company and has assets of over $600 million, making annual grants of around $27 million to most of the main environment groups: Wilderness Society, World Resources Institute, Sierra Club, Friends of the Earth, Earth Island Institute, Environmental Working Group, Worldwatch Institute, World Wide Fund for Nature, Natural Resources Defense Council, and Public Employees for Environmental Responsibility. Surdna and Andrus also have private forestry holdings, which have benefited from the efforts of environment groups to limit logging on public forests in the Pacific Northwest.

There are numerous other grant-giving foundations supporting environment groups, but Pew, Blue Moon and Surdna are three of the largest players. As well as the prominent environment groups, however, all three are substantial sponsors of three organizations of particular interest: the Union of Concerned Scientists, the Tides Center/Tides Foundation, and Environmental Media Services.

The Tides Foundation was established as a public charity in 1976, and receives and then directs funds to numerous causes, allowing donors to mask the groups they are funding. Exploiting a legal loophole, non-profit entities can set up for-profit entities and direct money to them through Tides, through 'donor-advised funds' which are for specified groups or specified purposes. The Pew Charitable Trusts, Tides largest single funding source, established three for-profit media companies (Pew Center for Excellence in Journalism, Pew Center for Civic Journalism and the Pew Center for the People and the Press) which are funded with money from Tides, which collects a standard 8 per cent fee from the money donated to it by Pew. These groups run public opinion polling and promote journalism to support the issues Pew considers important. (US tax law would prohibit Pew from funding such activities directly.) Tides also funds the Institute for Global Communications and the Independent Media Center.

Where no group exists, Tides can create a group to meet the identified 'need'. The Tides Center, established in 1979, is thought to provide a liability 'firewall' for the Tides Foundation from any possible legal action by those who might be harmed by Tides-funded activities, such as loggers put out of business by groups it funds. Tides funds groups such as the Ruckus Society (which was active in the protest against the WTO in Seattle in 1999), the Wildlands Project (which featured in the Lomborg controversy), the Wilderness Society, Natural Resources Defense Council, Environmental Working Group, Friends of the Earth, Greenpeace, Environmental Media Services and the Union of Concerned Scientists. The Natural Resources Defense Council, Environmental Working Group and Environmental Media Services began as projects incubated by the Tides Center. Environmental Media Services host the website *RealClimate* established by Mann and his supporters when the Hockey Stick ran into trouble.

Environmental Media Services was established by the Tides Center in 1994, became an independent non-profit organization in 2002, and is co-located in the same office suite as Fenton Communications, a public relations company that refers to itself as the leading public interest firm in the USA. Fenton Communications ran the 1989 campaign against Alar, a chemical which was used to help get apples to the market in better condition. Fenton ran a scare campaign starting with a *60 Minutes* programme creating fear that Alar was carcinogenic, and exploiting the celebrity of actress Meryl Streep and using groups which were incubated by the Tides Center: Mothers and Others for a Livable Planet and the Natural Resources Defense Council.

Fenton leaked to CBS an alarming report suggesting Alar was carcinogenic, especially to children, but the report was not peer-reviewed and was not based upon human exposure. Alar is not considered to have constituted a significant risk, and the case was a landmark in establishing the need for more cautious use of data for human chemical risk assessments. Fenton Communications has been associated with numerous other environmental and public health scares, including swordfish, breast implants, the dangers of pthalates in medical products, bovine growth hormone and the W. Alton Jones project with EDCs, including the publication of *Our Stolen Future*. Fenton Communications principal David Fenton was a founder of EMS, and a board member of the Environmental Working Group.

The president of EMS was Arlie Schardt, who worked on Al Gore's presidential campaigns in 1988 (as national press secretary) and 2000 (as communications director) and was also Senior Counsellor at Fenton. EMS provides experts to contribute material to the media (including sources at UCS, Greenpeace, NRDC and WWF), and acts as a not-for-profit group that can assist with Fenton's 'black marketing' – raising concerns about

BGH in milk, and thus (for two years from 1998) helping sell Fenton client Ben and Jerry's ice cream, which is labelled as 'hormone-free'. The Ben and Jerry's Foundation is in turn a donor to Tides, and Ben and Jerry's also donates 20 per cent of the profits from its 'Rainforest Crunch' ice cream to Tides. Ben and Jerry's had won a lawsuit in 1997 allowing it to claim that the absence of rBGH in its ice cream was an advantage for consumers, despite the Food and Drug Administration maintaining that there was no scientific test available that could distinguish between milk from rBGH treated cows and untreated cows.

One substantial donor to Tides has also been Teresa Heinz Kerry, wife of 2004 Democratic Party presidential candidate Senator John Kerry who, through the family's trusts, Heinz Family Foundation and Howard Heinz Endowment, has given over $8 million to Tides, securing the establishment of a Tides Center of Western Pennsylvania (Mrs Heinz Kerry is a Pittsburgh native). Heinz money supports the Heinz Environment Prize, which carries an award of some $250 000 and was once awarded to Dr James Hansen, who (despite being a public servant at NASA) campaigned for Kerry in the 2004 election in the key state of Iowa, and complained of being gagged in 2006 when NASA management would not let him engage in advocacy for climate change policies.

POLITICAL SCIENCE AND POLITICAL INFLUENCE

Donald Kennedy (2006a) wrote an editorial in *Science* in February 2006 lamenting the 'gagging' of James Hansen at NASA. He relied upon a report in the *New York Times* for the facts of the case, and lamented this (and the 'gagging' of NOAA scientists from speaking out in favour of a link between climate change and hurricane intensity) as 'part of a troublesome pattern to which the Bush administration has become addicted: Ignore evidence if it doesn't favour the preferred policy outcome.' But Hansen, who must possess a remarkable ability to speak while gagged, had made a press release claiming 2005 tied with 1998 as the warmest year on record, which was represented as a statement by NASA, when NASA had not authorized such a statement, and the circumstances suggest a case of a scientist on the public payroll who not only did not wish to be bound by the rules applying to public servants, but wished to make statements on behalf of his government agency.

Indeed, Hansen's Goddard Institute of Space Studies (GISS) within NASA had been unavailable to answer media requests about 2005 data, and had informed NASA public affairs that no press release by GISS was planned. NASA headquarters public affairs office provided Dr Waleed

Abdalati for an interview with the *Los Angeles Times*, and he could not confirm that 2005 tied for the hottest year on record. Nobody from GISS or the Goddard Space Flight Center (GSFC) was able to confirm any data to this effect. NASA Headquarters Public Affairs Officers responded to a query from ABC News in the same manner. ABC wanted to run something on their *Good Morning America* programme, and it hinted that it was in discussions with Hansen over the story. Neither Hansen nor GISS nor GSFC confirmed the release of any data with NASA public affairs, and it thought the press enquiries had been dealt with, and that no release of data would be made on December 15. Then Hansen released the data on December 15 in an interview given exclusively to ABC and in a letter to the editor submitted to *Science*, a release that had not (as required by NASA rules) been approved and coordinated by NASA headquarters.

The memorandum recording the NASA series of events concluded with the following paragraph:

> It bears mentioning that Dr Hansen and GISS as a whole do not always follow Headquarters protocol and sometimes fail to give Headquarters adequate notice when they plan to release information to the public. This is not the first time Headquarters has had this problem with GISS. (NASA, 2005)

It transpired that George Deutsch, the public affairs officer responsible for this 'gagging', had partisan connections with the Bush 2004 election campaign, and had misled NASA as to his credentials in his resumé (stating he had finished his degree, when he was one paper short of graduating). Was the 'gagging' politically motivated? Quite possibly, but Hansen not only broke with procedures (seemingly not for the first time), but was quite clearly cultivating the media by offering them a story with a political purpose, since part of his message was that the largely voluntary measures proposed by the Administration to address climate change were likely to prove inadequate. Given that Hansen had campaigned against Bush and for Kerry in 2004, it was hardly just the Administration that was politicizing science here.

But the science in question was also largely virtual. The statement that 2005 was the warmest on record (GISS) or second warmest (according to NCDC) depended upon *estimates* of Arctic temperatures, with James Hansen stating that 'the inclusion of estimated temperatures is the primary reason for our rank of 2005 of the warmest year' (Mercosur News Agency, 27 January 2006). The statement depended on filling in inconvenient data gaps with what were thought might be appropriate values – a step which should have given rise to caution in how the data was presented to the public.

There was a similar 'filling in' of data gaps with a NASA graphic of Arctic sea ice in 1979 and 2005 published on their website in 2006. Not only was sceptic Pat Michaels quick to point out that 1979 was a favourable year for comparison, coming at the end of the post-war cooling period, but an absence of data in the period 1979–85 was addressed by placing a solid white disc of ice polewards of Greenland (the sharp edges of which were visible in two places upon close inspection). This area may well have been solid ice, but to extrapolate ice into this area on the graphic rather than mark it as having insufficient data lent a false authority to the picture of Arctic sea ice (see the cover illustration).

Science editor Kennedy made no apparent attempt to discover the facts behind the 'gag' story, and used it to criticize the Bush Administration. And in his treatment of the other supposed instance of 'gagging' he revealed a preparedness to use remarkably lax standards of science to support his pre-existing views. Citing MIT scientist Kerry Emanuel's research suggesting hurricane intensity had increased with ocean surface temperatures over the past 30 years, Kennedy bemoaned the NOAA prohibition on their staff speaking to reporters or presenting papers at conferences without departmental review and approval – again, a perfectly normal constraint at public agencies and even at some seemingly independent research institutes.[3] Kennedy noted that the NOAA website attributed increased hurricane strength to 'tropical multidecadal signals' affecting climate variability, for which a 30-year analysis could hardly provide a refutation, yet Kennedy thought that the matter had been settled by Emanuel providing 'convincing evidence against it in recent seminars'. Here was the editor of one of the most prestigious scientific journals revealing a substantial bias, and expecting us to accept seminar papers (not yet subjected to any peer review) as authoritative evidence.

At almost the same time as Kennedy was accepting Emanuel's seminars as scientific gospel, a World Meteorological Organization Steering Committee (the membership of which included Emanuel) was unable to reach a conclusion on the question of rising intensiveness, while finding no evidence of increasing frequency and attributing increased damage costs and disruption to increasing coastal populations and rising insured values (WMO, 2006).

This use by Kennedy of seminar presentations as sources of authority, together with the filling in of data gaps in ice and temperature records by NASA, do not strictly constitute outright dishonesty, because Kennedy sincerely believed Emanuel to be correct, and NASA assumed (perhaps correctly) that its extrapolations to fill in the gaps were justified by what was probably the case. But these are examples of a phenomenon that can only be described with an impolite word: bullshit. Bullshit stems from its practitioners seeking to convey a greater degree of knowledge than they possess in order to exercise power, since bullshitting bestows a greater degree of authority on the practitioner, and authority is a form of power. Harry Frankfurt (2005) argued that bullshit was the inevitable product of public life, where people were driven by their own motivations or the demands of others to speak extensively on matters of which they are to some degree ignorant (see also Penny, 2006). The *honest* response to a lack of data would be to make the deficit explicit (or, with uncertain science, the disagreement), but that would limit the ability to make authoritative statements, and it is this desire to speak with authority, with power, that leads to the situation where power speaks to truth, rather than the reverse.

It is this extending the evidence, this 'bullshitting', rather than explicit dishonesty that best characterizes the nature of virtuous corruption, because it is the virtual or incomplete character of the science which permits honest people to go beyond what a strict adherence to the science would permit, and it is the desire to speak with authority, to exercise power, in a good cause that motivates them when they really should remain silent. Unfortunately, in the environmental sciences, we find both a lack of commitment to transparency and disclosure (as we saw with the Hockey Stick case) coupled with a willingness to gloss over or even accept inaccuracies if they point in the 'right' direction.

For example, a paper in *Nature* using the species–area model to predict species distribution in response to modelled climate change (in turn based upon emissions scenarios) concluded its abstract with a call to action (Thomas *et al.*, 2004, p. 145): 'These estimates show the importance of rapid implementation of technologies to decrease greenhouse gas emissions and strategies for carbon sequestration.' No evidence presented in the paper supported this proposition. Yet this paper, predicting that a million species might eventually become extinct as a result of climate change over the next 50 years, was based on an analysis of only 1103 species, an extrapolation that was exaggerated by both the media and EU Environment Commissioner Margot Wallstrom, who stated that a third of all species could be wiped out by 2050, ignoring the qualifier 'eventual' and the fact that 2050 referred only to warming until that point, not the 'projected' extinction number by that time (Ladle *et al.*, 2004). But the nobility of the cause was thought by some to outweigh any concern over accuracy, with one communication in *Nature* being titled 'Extinction-risk coverage is worth inaccuracies' (Hannah and Phillips, 2004), and another scholar stating that 'the major usefulness of such exercises is in destroying any residual complacency about climate change among conservationists and hopefully among policy makers, and in highlighting that a conservation strategy based upon a static and isolated set of parks is unlikely to be robust to global climate changes' (Lewis, 2006, p. 170). That is to say, it is acceptable to bullshit as long as it is done in a good cause.

This was not an isolated case. In January 2006, *Nature* published a paper linking frog decline in Central and South America to global warming (Pounds *et al.*, 2006). In fact, the extinctions were caused by the introduction of the chytrid fungus in the period 1984–96 (La Marca *et al.*, 2005); Pounds *et al.* were trying to link the spread of the fungus to the (very small) warming observed in the region, and had earlier linked frog decline to the decreasing frequency of mist (Pounds *et al.*, 1999). They seemed keen to link frog decline to climate change in more than one way, and provided evidence of their attachment to the political cause (amazingly) in the abstract to their 2006 paper, where they stated 'With climate change promoting infections, disease and eroding biodiversity, the urgency of reducing greenhouse gas concentrations is now undeniable.' Again, this was merely asserted; there was no reason or evidence presented in the paper to support the rhetorical device that the need for urgent mitigation was 'undeniable'. (In fact, the introduction of the fungus was clearly the primary cause of decline, and it probably resulted from human agents – possibly scientists or ecotourists – but there was no call for restrictions on either of these.) *Nature* now permits not only the framing of science in 'noble' political contexts, but allows this in a summary of papers which do not address such dimensions and provide no evidence.

Some scientists are concerned at the exaggeration of the science on issues like climate change. Hans von Storch, a believer in climate change, put it this way to the BBC in April 2006: 'The alarmists think that climate change is something extremely dangerous, extremely bad and that overselling a little bit, if it serves a good purpose, is not that bad' (Cox and Vadon, 2006). The media is undoubtedly complicit in this exaggeration, but scientists themselves seeking headlines tend to allow the exaggeration and few ever make any effort to have the exaggeration corrected. For example, in 2005 Climateprediction.net ran a simulation of climate using the unused capacity on people's networked PCs that was published in *Nature* despite having more to do with public participation than original science. Most of the results were unremarkable, but some models produced a barely credible 11°C for a doubling of CO_2. Predictably, this is the figure the media headlined.

Dr Myles Allen, principal investigator at Climateprediction.net, blamed the media, stating 'If journalists decide to embroider on a press release without referring to the paper which the press release is about, then that's really the journalists' problem. We can't as scientists guard against that' (Cox and Vadon, 2006). But this is disingenuous. It is highly unlikely that any journalist will have the time (or the skills) to read a scientific paper critically – which is why scientists provide them with press releases. And this particular press release was shown to several climate scientists by the BBC, which reported that 'All were critical of the prominence given to the prediction that the world could heat by up to 11°C' (Cox and Vadon, 2006).

But government agencies can also add to the hype. The UK Environment Agency publicized research on climate change over the next 1000 years and predicted cataclysmic change (temperature rises of 15°C and sea level rise of 11 metres) while arguing that action was needed urgently. But the study's author, Dr Tim Lenton criticized the presentation of their science by the Environment Agency, pointing out that if nothing were done for a century you would still only get a fraction of the worst-case scenario. Clive Bates, head of environment policy at the Environment Agency, considered that Dr Lenton simply failed to understand the way the media worked. He stated that 'He was involved in signing off on the press release, there is nothing in there that is actually incorrect.'

This exaggeration by the media, as well as that by green groups such as Greenpeace, was also criticized by the Royal Society in a leaked confidential memo reported in the *Guardian*. Predictably, however, the newspaper gave prominence to the Royal Society's concern that 'groups and individuals' might attempt to question the science of global warming and the need to curb greenhouse gas emissions, giving the story the headline 'Scientists fear new attempts to undermine climate action' (Adam, 2006). When even the Royal Society is more concerned with politics rather than science, the problem of the politicization of science is clearly widespread, and some have begun to remark on this. In 2005 *The Lancet* declared that the Royal Society had ceased to be 'a place to discuss the subversive subject of science' and had become 'self-serving and parochial'. (*The Lancet*, 2005). This state of affairs occurred during the term of Lord May as president, and is perhaps not surprising given his longstanding commitment to the cause.

A young Lord May wrote a neo-Malthusian survey of the environmental crisis in 1971. Interestingly, he saw the threat of climate change coming not from an accumulation of GHGs, but from the release of heat generated by man's activities, with the 'climatological heat limit' of 1 per cent of the solar energy absorbed and re-radiated likely to be reached in less than a century (May, 1971, p. 124). May's paper referred to almost no social science literature and demonstrated the kind of naïvety social scientists see only too often from natural scientists who wander into the social realm: 'were the population to continue to increase indefinitely at its current rate, then in 400 years there would be one square yard for each inhabitant of the globe', he opined (May, 1971, p. 123). He also surveyed resources in an analysis largely bereft of economics, but his most remarkable and statement was perhaps his suggestion that studies of overcrowding among rats could tell us something about human behaviour:

> Even though abundantly supplied with food and places to live, overcrowded rat communities provide a spectacle of social chaos, with, inter alia, complete disruption of maternal behaviour, sexual deviations including homosexuality, hyperactive and totally withdrawn individuals: in short, all the forms of aberrant behaviour one finds in say, New York City. (May, 1971, p. 124)

Lord May even committed the genetic fallacy in his Anniversary Address as he stepped down as President of the Royal Society, asserting (without producing a single piece of evidence):

> Not surprisingly, there exists a climate change 'denial lobby', funded to the tune of tens of millions of dollars by sectors of the hydrocarbon industry, and highly influential in some countries. This lobby has understandable similarities, in attitudes and tactics, to the tobacco lobby that continues to deny smoking causes lung cancer, or the curious lobby denying that HIV causes AIDS. (2005, p. 8)

This is a simplistic view of corporate influence and science. Gas companies stand to make money from decarbonisation policies, because natural gas provides the cheapest and easiest alternative to coal. Similarly, multinational chemical companies were enthusiastic supporters of a phase-out of DDT in developing countries during the negotiation of the Stockholm Convention on Persistent Organic Pollutants, because it was out of patent and there was benefit in a policy that advantaged their more expensive patented alternatives. The political Right might have opposed the proposed ban on DDT, but it was not out of support for corporate interests in opposing good science. Indeed, it was environmental activists and regulators who abused the science on DDT.

The banning of DDT is regarded by some as a case of scientific fraud, because many of the effects attributed to the chemical are supported by weak evidence at best (Edwards, 2004). For example, the most notorious putative effect of DTT was it causing the near extinction of bald eagles and peregrine falcons by thinning their eggshells as a result of biomagnification up through the food chain. Yet bald eagles were threatened with extinction in the lower 48 US states as early as the 1920s, and peregrine falcons were reduced to 170 breeding pairs in the Eastern USA by 1940. DDT was not manufactured anywhere until 1943, and while a paper by Bitman *et al.* (1970) published in *Science* reported thinning of shells with DDT exposure and reduced levels of dietary calcium, *Science* refused to publish the subsequent finding that shells were not thinned by DDT exposure when there was adequate calcium. Instead, Bitman *et al.* had to publish in an obscure specialty journal (Bitman *et al.*, 1971) and many continued to believe that DDT caused egg thinning.

DDT was banned not because of any environmental effects, but because it was judged by US Environmental Protection Agency (EPA) Administrator William Ruckelshaus to be a human carcinogen. An extensive review by the EPA in 1972 concluded that DDT was not a carcinogenic hazard for man, yet Ruckelshaus banned it two months later on the basis of two questionable animal studies, whereas human epidemiological studies revealed no elevated incidence of cancer. The decision, which appeared to be driven by the political effect of Rachel Carson's book *Silent Spring*, especially its fictional scenario of a world with no birds (the 'silent spring'), applied not only to domestic use in the USA, but was imposed (through conditions attached to US aid) on developing nations, where the ban has caused millions of deaths (perhaps 30 million) through malaria and other diseases with insect vectors that could have been controlled with residual spraying of walls with DDT. The Stockholm Convention on Persistent Organic Pollutants was to exempt DDT in developing countries from a phase-out, and in 2006 the WHO finally approved residual spraying of DDT. It was politicized science that produced this result, and at a time long before the 'sound science' initiative.

CONCLUSION

We are familiar with the possibility that financial factors might have a corrupting influence on science. Lucrative research contracts can lead to lapses in procedure in the conduct of science, and regulators have responded by imposing guidelines such as those of Good Laboratory Practice – agreed internationally by the OECD – and regimes of auditing to minimize the incidence or error and fraud. And chemical corporations have sometimes tried to suppress unwanted research (Weiner, 2005). It is easy to point an accusatory finger at the influence of money in regulatory science (Beder, 1997), but the system nonetheless works reasonably well, and we can even point to some results which are highly improbable if one accepts the conspiracy theory approach to regulatory science and corporate influence. For example, the carcinogenic effects of TCDD dioxin were found by a team of scientists employed by Dow Chemical, hardly a result in the interests of the corporation. (The same paper first reported observations of hormesis with dioxins, because cancer rates were *lower* with low exposures, such as were actually likely to be experienced.) Moreover, the financial interests at stake with climate change and chemicals are relatively obvious; the noble cause corruption of environmental science is perhaps the more pernicious for being much less obvious. Regardless, *both* can serve to undermine good policy-making on matters relating to the natural environment and human health.

It is within this context that we must view the claims by the Union of Concerned Scientists that the Bush Administration was politicizing science by insisting on such measures as the requirement for science informing regulations to be peer-reviewed. The Bush Administration inherited an executive branch which had been extensively populated not just by partisans, but by those closely associated with environmental groups, and the Union of Concerned Scientists itself is closely associated with those same groups. (There is a need, perhaps, for a Union of Disinterested Scientists.) The virtuous corruption of science has been very well funded, but the main concern here is with the significance of the cause in bringing about that corruption. We have suggested here that this is facilitated by the virtual nature of much of the science, where scientists are likely to want to fill in the gaps – to 'bullshit' – in order to speak with greater authority than is justified in support of a good cause.

But there remains something of a puzzle, and it is this: there seems to be a closer affinity between environmentalism and Left-leaning parties in western democracies, and greater hostility towards environmental protection from Right-leaning parties. It remains for us to explain why there is this connection and how it impacts upon the corruption of science, and this is the task of the final chapter. There, the argument will be developed that there are cultural factors associated with appreciations of nature that align with relevant political ideologies, and these 'cultural filters' are more important when the science involved is more virtual and less reliant upon clear observational data.

NOTES

1. *General Electric Co. v. Joiner*, 1997, 522, US 136, 118 S. Ct. 512; 139 L.ed.2d 508.
2. *Kumho Tire Company, Ltd., et al., v. Patrick Carmichael, et al.*, Supreme Court of the United States No. 97-1709 argued December 7, 1998 – decided March 13, 1999.
3. The author is associated with the Antarctic Climate and Ecosytems Cooperative Research Centre and is subject to a similar constraint, as are all its staff.

6. Science and its social and political context

He who knows only his own side of the case knows little of that.

John Stuart Mill

Perhaps the best-known modern example of how politics can contaminate the conduct of science was Soviet geneticist Trofim Lysenko's rejection of the 'dangerous' Western concepts of Mendelian and Darwinian genetics and evolution in favour of somewhat bizarre Lamarckian views that, under a socialist system, cows could be trained to give more milk and their offspring would then inherit these traits. Heisenberg's uncertainty principle had received similarly short shrift in Soviet science. Claus and Bolander have noted the key features of what is now known as Lysenkoism, and many can be seen in the politicized science of today: a necessity to demonstrate the practical relevance of science to the needs of society; the amassing of evidence as substitute for causal proof as the means of demonstrating the 'correctness' of the hypotheses; ideological zeal supplanting devotion to science, so that dissidents could be silenced as enemies of the truth. Manipulating data to support the ideological cause was permissible, since this was a higher truth (Claus and Bolander, 1977; see also Cole, 1983). But a close relationship between politics and science is something on which the political Left does not have a monopoly.

Lysenkoism might appear to be a rather extreme example of social and political factors influencing the conduct of science, but there is ample evidence that much of the science relating to environmental problems is at least at risk of being contaminated by similar influences. There is a strong case that what is broadly known as 'environmental science' is particularly prone to the intrusion of various non-scientific concepts including the ideology of environmentalism. While many scientists advocating particular policy measures might think that the 'science' leads them to espouse particular political positions, their values and politics are frequently driving their science, and certainly their environmentalism. For this reason it is worth examining the underpinnings of environmentalism in some detail.

In this final chapter I explore some lines of explanation for the virtuous corruption of virtual science, including its relationship with the political

Left. I suggest the answer lies in the (inevitable) intrusion of what we might refer to broadly as 'cultural factors', which are of greater importance where the science is less determinate – where there is reliance upon data which must be substantially manipulated in order to be used, and where there is greater reliance upon mathematical modelling rather than observational science. The chapter first shows that the proposition that environmentalism entails some kind of progressive political ideology is false, and that ecology has a 'dark side' of possibilities of which we need to be aware (and wary). I then show that such factors pervade environmental science (especially conservation biology) before concluding by showing that the literature on the cultural theory of risk can provide some explanation of the tendency observed here for progressive politics to encourage a greater degree of noble cause corruption: while it can happen on both sides of politics, and no particular political ideology is entailed by environmentalism, the political Left shows a greater proclivity than the Right, and must raise its expected standards of environmental science if it is to put policy-making on a firmer evidence base (and save science from the political 'war' it wrongly perceives as having been initiated by the Right).

ECOLOGY'S DARKER SIDE

Scientists, despite their frequent protestations to the contrary, never rise completely above cultural factors. Scientists like to think that they work solely on the basis of reason and evidence, but they are always to a greater or lesser extent wrong. Environmentalism involves both a set of values *and* a body of scientific knowledge, and the two appear to be connected. Some value contexts are more conducive to certain scientific conclusions than others, though not always in predictable or even desirable ways. If one reads the statements of the Union of Concerned Scientists, one could be forgiven for thinking that environmental concern and liberal Left values are inextricably linked, that one entails the other, but this could not be further from the truth. Science operates within a context where funding decisions, career opportunities, cultural values, strategic imperatives, and many other factors influence its conduct. This is not to say that the results produced by science are always debased by such factors, but to alert us to the possibility that they might be. The radical theories in genetics of Trofim Lysenko were certainly wrong, but they resonated with the ideology of the Soviet state.

The theories developed by scientists frequently reflect certain social phenomena, and this is certainly the case with ecological science. The very notion of sustainability came from forestry in Germany and reflected not just concern with the sustainability of harvest levels, but concern with social stability in uncertain times. The very word 'ecology' came out of the same context, termed by Ernst Haeckel in Germany in 1866, and the concept of an ecosystem, at first embodying assumptions (now rejected) of harmony and self-regulation, was developed in the context of a very uncertain environment of Europe in 1935, being developed in Britain by botanist A.G. Tansley. As noted earlier, ecology is also full of concepts like 'alien', 'natural enemy' and 'invasion' – concepts more at home in the Europe of the 1930s than now (Chew and Laubichler, 2003).

Germany, in fact, provides us with a significant counterfactual to the suggestion that environmentalism merely reflects affluence, rather than (or at least additionally) more deeply seated cultural factors. It also provides a counterfactual to the notion that environmentalism necessarily entails a progressive political agenda. Germany adopted an extensive green agenda in the absence of affluence – indeed, at a time of substantial economic insecurity for its citizens. In the face of economic insecurity, the German government once adopted restrictions on suspected carcinogenic food additives, on asbestos, and on suspected carcinogenic pesticides. It mounted vigorous campaigns against cigarette smoking, adopted regulations requiring whole grain to be used in bread, expressed official concern about vivisection and gave encouragement to organic agriculture (Proctor, 1999). This was not a contemporary German government responding to the emergence of the Greens as a political force but the government of the Nazi era, which drew strength from a period of decided economic *insecurity*. Affluence cannot provide much of an explanation for the establishment of an organic garden at Dachau, nor the fact that there was even a native landscape aesthetic in the landscaping of Nazi-era autobahns (Rollins, 1995). We have to turn to other factors for an explanation – cultural factors, historical factors – which favoured the adoption of such a green agenda in Germany in the 1930s.

Anna Bramwell and others have explored the significance of ecological thought in Nazi Germany, especially the 'blood and soil' movement led by Rudolph Darré. Many critiques of the darker corners of the environmental movement (both focused and more incidental) have made this point: Stephen Budiansky's *Nature's Keepers*; Alston Chase's *In a Dark Wood*; Simon Schama's *Landscape and Memory*. These have not just been right-wing critiques seeking to undermine the new, 'progressive, green left'. They have been perhaps most severe when they have come from the Left, such as Janet Biehl and Peter Staudenmaier's *Ecofascism: Lessons from the German Experience* (1996). By combining in the one volume two essays ('Fascist Ideology: The Green Wing of the Nazi Party and its Historical Antecedents' and 'Ecology and the Modernization of Fascism in the German Ultra-Right') these critics from the Left attempt quite explicitly to link those neo-fascist elements within modern Green politics in Germany (which the left wing of the party had to purge) with their Nazi antecedents. All this is quite clearly done to suggest that modern environmentalism has fascist leanings, but this is taking the argument too far. While there are dark possibilities, environmentalism need not either entail fascism, nor lead to it.

Modern environmentalists would nevertheless do well to ponder where a rejection of humanist values and a belief in a transcendental ecocentric value system might lead them, just as there is value in studying the resonance between the Nazi war on cancer and their wars on communism and Judaism, which were included in their cancer metaphor. In this regard, Ehrlich's invocation (1968, p. 152) of the metaphor of cancer to describe human population growth in *The Population Bomb* is somewhat disconcerting. But it is simplistic in the extreme to suggest that environmentalism leads to the gates of Auschwitz. It is equally mistaken however, to assume that environmentalism somehow *precludes* such a path and leads instead to some progressive future. There are indeed some worrying connections between ecologism and various extreme political views, not least of which are the celebrated racist views of Haeckel, the father of ecological science. It is more plausible to suggest that there are many complex factors at work. Indeed, Nazism and ecofacism or other darker contemporary manifestations of ecological thought are better seen as *reflective* of similar underlying factors in history and culture.

Both might also be seen as millenarian movements, for example. This is quite clear with Nazism and its 'Thousand Year Reich', but there is a similar appeal to a stable utopian end point after apocalyptic collapse in much environmentalism (Gelber and Cook, 1990). The neo-Malthusian spectre of the Club of Rome was one of rapid, catastrophic growth in population and economy, demanding 'zero population growth' and 'zero economic growth' or (in its more sophisticated forms) a 'stable state society'. The origins of the word 'sustainability' – 'that magic word of consensus' as Worster (1993, p. 144) puts it – lie in the concept of 'sustained yield' which emerged first in scientific forestry in Germany in the late eighteenth century. As Robert Lee (cited by Worster, 1993, p. 145) has noted, it came not just as a response to the decline in German forests, but as a response to the uncertainty and social instability which wracked Germany at that time (and which were responsible at least in part for the decline in German forests). It was an instrument of a strong state for ordering social and economic conditions which stood as a 'necessary' counterweight to emergent laissez-faire capitalism.

There is a long tradition of Western thought involving decline, often catastrophic decline, from some idyllic past – usually as a result of some sin or degeneration. Hanson (2006) lists Hesiod (the eighth-century BC Greek poet), Sophocles, Virgil, Ovid, the Biblical Fall, and numerous others through Rousseau, Nietzsche and Spengler to Rachel Carson, Paul Ehrlich and Jared Diamond. Diamond, author of two hugely popular books in *Guns, Germs and Steel* and *Collapse*, manages in one to attribute the success of the West, not to freedom, rationalism, individualism, and consensual government, but to inanimate, natural forces at work; and in the other, the forthcoming collapse is predicted on the basis of a few atypical examples (Easter Island, Greenland, Pitcairn Island, and so on) because the failed societies degraded their environments through ignorance and greed and thus (seemingly deservedly) disappeared.

What is telling is that environmental activists, most social scientists writing about environmental issues, and many 'activist' environmental scientists still cling to the myth of the 'balance of nature' that has long been rejected by ecology (Scoones, 1999). By accepting this myth in the face of the scientific evidence, any change in ecosystems or climate can be attributed to human agency, and imparted with deep social meaning – either apocalyptic or (if promising some eventual return to a stable state of grace) millenarian. Regardless, Hanson suggests that such pessimism fulfils a need in affluent but guilty Westerners to feel bad about their privileges without having to give them up.

There have been numerous analyses of environmentalism in millenarian terms. For example, Buell (1995) has analysed 'environmental apocalypticism', while Killingsworth and Palmer (1996) and Lee (1997) have described the millenarian aspects of the contemporary environment movement. Stewart and Harding saw environmental concerns as but one of a number of *fin de siècle* concerns:

> During the 1990s, apocalypticism, and, somewhat less flamboyantly, its millennialist twin, have become a constant and unavoidable presence in everyday life. Idioms of risk, trauma, threat, catastrophe, conspiracy, victimization, surveillance; social, moral, and environmental degradation; recovert, redemption, the New Age, and the New World Order permeate the airways. (1999, pp. 289–90)

Stewart and Harding also point to attributes of apocalypticism that describe the constant ascribing of sinister motives to those who present dissenting views: climate sceptics are in the pay of fossil fuel corporations, Lomborg's analysis will assist these interests, and so on. Such conspiracy theories at once serve to defend the prevailing paradigm, reinforce solidarity among the adherents and reinforce their sense of purpose:

> Conspiracy theories can identify absolute truths about the world while dismissing holders of power as sinister, corrupt, and deceptive; they can also resurrect agency and the sense of a privileged community 'in the know,' and an otherwise bleak present can become charged with purpose and focus. (Stewart and Harding, 1999, p. 294)

Scientists such as Lord May who commit the genetic fallacy, attributing the dissident views of climate sceptics to the 'sinister, corrupt and deceptive' antics of ExxonMobil, and who make statements that are logically identical to accusations of witchcraft, probably do not think that they have much in common with pre-Enlightenment societies. But the mixture of climate change and witchcraft is not a new one (Behringer, 1995), and there is no reason to suppose that scientists are above the defensive psychology Festinger (1962) termed 'cognitive dissonance' – coincidentally developed in an earlier study of the state of denial found in millenarian movements when their prophesies failed to materialize (Festinger, *et al.*, 1964). But May (and others) have agendas, are 'concerned' scientists, and no matter how much they consider they operate cognitively, this brings an 'affective' dimension to their thinking and they cannot rise above the same psychology all humanity exhibits. It is only scepticism and criticism that limits the extent to which affective factors intrude into science, and it is the nobility of the cause and thus the availability of strong moral arguments to disarm critics which facilitates it.

Nazism and environmentalism are by no means the only movements in the Western mainstream which can be analysed in millenarian terms. Both Nazism and environmentalism may well both contain elements of the romanticism Jeffrey Herf (1984) called 'reactionary modernism' but this is not to say that either necessarily entails or leads to the other. Two other movements which can be subjected to a millenarian analysis are Marxism (which promises an unchanging communist utopia after a period of revolutionary upheaval) and Christianity, especially in those manifestations which emphasize the heavily millenarian Book of Revelation. It is facile to suggest that either entails or leads to the other, and it is similarly facile to suppose that environmentalism entails or leads to Nazism – though (as noted above) it is wise to exercise caution about some of its darker possibilities. But this analysis shows how wrong it is to assume that environmentalism somehow entails a liberal democratic political philosophy, or the social democratic ideology that is described as 'liberal' in the United States. Classical liberalism, with its emphasis on the separation between the individual and the state, nevertheless provides a protection against the darker possibilities of environmentalism.

Culture and language might reflect and reinforce deep-seated cultural differences in responses to environmental threats: German environmental language frequently not only increases the threat image ('Klimakatastrophe') but also promises more control and salvation via state action ('Klimaschutz') (Boehmer-Christiansen, 1988). Steven Kelman has identified some aspects of Swedish political culture which he considered helped explain its position at the environmental vanguard. Sweden has a somewhat circumspect political culture, institutions described as accommodationist, and a tradition of the 'overhet' state: 'The people were expected to accept the notion of the good that the rulers defined.' To illustrate the point, Kelman quotes the words of an eighteenth-century Swedish poet inscribed over the entrance to the main hall of Uppsala University: 'To think freely is great, to think correctly is greater' (Kelmann, 1981, p. 121).

Culture, of course is a holistic concept, and political cultures cannot be readily disaggregated into neat constituent parts. Language, religion and attitudes are simply three facets of political culture. What is of interest here is the possibility that such cultural factors could exert an influence on the conduct of science. There is a range of possibilities in environmental concern, some 'progressive', some much darker. Environmental science does not lead ineluctably to 'progressive' political outcomes, and there are dangers in assuming that it does.

Science *can* be affected by values and interests, and the seeming impossibility of achieving 'objective' knowledge of any phenomena – natural or social – is sometimes taken to enfranchise a kind of 'anything goes' approach to knowledge often found in postmodernist texts, and endorsed by Feyerabend, sometimes referred to as 'postnormal' science (Funtowicz and Ravetz, 1993). All attempts to develop knowledge are, according to this view, 'just texts'. Any text becomes as 'valid' as another. But it does *not* follow logically that contamination of science by values and interests means we should assign it all some kind of equivalence. To draw attention to the pernicious effect of the extreme relativism of postmodernism on the use of science in the progressive cause, Alan Sokal (1996) submitted and had published as a genuine paper, a satire on postmodernist philosophy of science that mocked the idea of reality being inherently something of which we are unable to derive objective knowledge (as opposed to our understanding of it being always at risk of error and social construction). Sokal saw himself as a member of the progressive Left and lamented what had become of the role of science as a progressive force under the postmodernist enthusiasm that was sweeping the humanities. (He invited those who thought that science was a mere construction to test the Law of Gravity from the window of his high-rise office at Columbia University.)

We do have canons which help us tell (albeit imperfectly) good science from bad. Insistence on consistency of argument and adherence to the scientific method has brought numerous advances, not just in knowledge, but also in human welfare, with improvements in life expectancy, for example, that are more than 'just text'. And (though the pitfalls might be larger) the same holds for the social sciences: for all its limitations, economics plays its part in improving welfare; political science helps improve the making of public policy decisions. Indeed, the great advances in life expectancy for ordinary people during modernity came about through the design and financing of great public works projects such as the sewering of Victorian London (ironically, based on the erroneous miasma theory of disease) rather than on advances in treatments based on medical science. Contemporary environmentalism too often entails a rejection of modernity, rationalism and the Enlightenment. These attributes are not just found in environmentalism as an ideology, but they are deeply ingrained within ecological science itself.

ECOLOGY AND SCIENCE

Chase (1987) has explored the origins of ecosystems science, especially the extent to which it borrowed from physics a model that explained energy flows through a system, with nature operating much like a thermostat, so that ecosystems were seen as self-regulating and tending towards equilibrium. 'An ecosystem could be pictured in the form of a model like an electrical circuit, whereby energy, derived from the sun, was transferred by means of chemical processes through soil and grass up the food chain, and then, by decomposition, through the cycle again' (Chase, 1987, p. 313). Ecology lacked a scientifically respectable method for studying life, and the ecosystem approach provided scientific respectability by supplying ecologists with mathematical tools developed by physicists. It gave them access to the laws of thermodynamics and mathematics. Community ecology boomed, but the problem was 'true ecosystems . . . were hard to find. In fact, as even Tansley had acknowledged, they did not exist!' (Chase, 1987, p. 315). An ecosystem is nothing more than a construction, a tool by which scientists could artificially separate their subject of study from everything else. Early ecologists tried to study ponds as examples of ecosystems, but soon found that even ponds were not closed systems, since they were connected to the water table, and affected by underground currents and spring run-off. Migrating waterfowl both used them for food and deposited waste into them, trace elements of chemicals and nutrients fell into them in rain and airborne spores, and the sun provided energy inputs. When everything was connected to everything else on the globe, it was impossible to isolate an ecosystem to study, and even earth was affected by sunspots, meteor showers and cosmic radiation.

Leopold tried to develop his 'land ethic' by taking the obvious point that man was a part of ecosystems to something of an extreme. Whereas Tansley had taken this point to suggest that any separation between man and nature was artificial, and therefore that any human actions were just part of the system, Leopold took it to mean that man should develop an ethical system which included soils, water, plants and animals, or (taken together) the land. As we noted earlier (p. 32), environmentalists took to the idea of a self-regulating ecosystem with alacrity and the 'balance of nature' was granted sacred status (See Suzuki and McConnell, 1997).

The progress of ecological science was to diverge from this notion of a sacred balance, as change, perturbation and succession were seen to be accepted as core concepts. Jeff Harvey (2001, p. 463), Lomborg's harsh critic, still considers that 'healthy' ecosystems 'remain stable over comparatively long periods of time'. But the increased emphasis on mathematics which lent ecology its scientific gravitas helped steer it towards virtual science rather than experimental science. The emphasis on mathematical models de-emphasized the need for experiment, and the need for observational evidence. As we noted in Chapter 2, Chase puts it, 'It became perhaps too abstract, a discipline attracting deskbound number crunchers more than those who liked to tramp about the woods in wool shirts counting deer scat' (Chase, 1987, p. 322). Community ecologists reflecting critically on their discipline concluded it had too often been content with generalized mathematical theory and passively (rather than experimentally) collected observations (Chase, 1987, pp. 322–3).

The shift of environmentalism on to a religious plane, coupled with the descent of much of ecology in to the virtual world of mathematical modelling has facilitated the virtuous corruption of virtual science. That shift to a religious plane has often been to religions other than Christianity, and it has also been accompanied by the marriage of science to political activism.

There is no better example of the marriage of ecological science and activism than in the formation of the new 'synthetic discipline' of conservation biology in the mid-1980s by Michael Soulé, a former graduate student of Paul Ehrlich's at Stanford. Soulé turned from his post as a professor of biology to life in a Zen Bhuddism centre in 1978, but then stopped meditating and re-entered academia in the mid-1980s to found the new discipline of conservation biology, to which E.O. Wilson, Paul Ehrlich, and catastrophist Jared Diamond were early converts (Jones, 2003).

Soulé founded the Society for Conservation Biology in 1985 and its journal *Conservation Biology* in 1987. The Society has around 6000 members. Conservation biology was founded unashamedly as activist science. Soulé described it thus:

> Conservation biology, a new stage in the application of science to conservation problems, addresses the biology of species, communities, and ecosystems that are perturbed, either directly or indirectly, by human activities or other agents. Its goal is to provide principles and tools for preserving biological diversity. (1985, p. 727)

Soulé maintained that ethical norms were a genuine part of conservation biology, and with the spread of the 'discipline' to more than 80 universities, the basis for the noble cause corruption of ecological science was laid. As one writer put it, Soulé combined 'very solid ecological science with a zeal for connecting that science with conservation' (Jones, 2003). His activism included teaming up with Dave Foreman, the founder of the radical biocentric group Earth First! (famous for its monkey-wrenching, tree-spiking and other radical actions, and featuring a millenarian ideology (Lee, 1997)) to establish the Wildlands Project to link vast areas of North America for ecosystem protection and, particularly, to reintroduce large carnivores which Soulé regards as 'the governors of ecosystems' without which ecosystems will collapse (Jones, 2003). (Recall that the endorsement of the Wildlands Project by Paul Ehrlich and E.O. Wilson was an issue between Lomborg and critics.)

Soulé confesses to both a political purpose and preference for the emotional commitment to that cause over hard science. He has been quoted as saying:

> I once wrote that the facts compute, but they don't convert. I know when I'm giving a lecture and tears come to my eyes it has much greater impact than slide after slide of numbers, or even pretty pictures. An instant of honesty and compassion is much more important than an hour of logical argumentation and the facts. (Jones, 2003)

This preference for emotions over facts is fortunate, because the problem with the science of conservation biology is that it cannot answer basic questions, such as (even approximately) how many species there are on earth, and it really can provide no reliable estimate of the numbers of species that are becoming extinct, so it has no chance of providing credible information about rates of extinction. Yet we have an international agreement entailing a commitment to addressing that which cannot be measured: the Parties to the Convention on Biodiversity agreed in 2002 to reduce the rate at which animals and plants were disappearing by 2010. As Jeff McNeely, chief scientist at IUCN-the World Conservation Union, acknowledged on the eve of the Eighth Conference of the Parties to the CBD, 'The implication of not knowing exactly how many species there are is that we can't tell we are actually making progress on the 2010 target' (Reuters, 2006). So we have the somewhat farcical situation that the false concreteness of estimates of species extinction based on virtual science have succeeded in driving parties to reach international agreement to do something that they cannot measure, so they cannot tell whether they have succeeded or failed.

McNeely acknowledged the problem. Only around 1.7 million plant and animal species had been described, but some estimates were that there might be 100 million species. And the gulf between the numerous estimates of extinction numbers (unable to be described as rates without an estimate of total numbers) and actual documented extinctions was even wider, with the WCU estimating only 'more than 800' plant and animal extinctions since 1500 when accurate historical and scientific records began. But the magnitude of the problem depends more on human valuation than raw numbers. Even if we were to accept the 'virtual extinctions', it is our evaluation that this is a 'dangerous' or 'unacceptable' number, and that it matters that human agency is the cause that converts the data into a problem. Science cannot establish either of these. One estimate is that there have been somewhere between 5 and 50 billion species alive on earth over its history (Raup, 1991). Only about 1 in 1000 species that have ever lived are still alive, making for an enormous failure rate as part of evolution. On the basis of these data, even if the current rate is higher than the 'natural' rate, it is not clear that it is either alarmingly high or necessarily a bad thing.

As noted earlier, ecology involves all manner of projections of human values on to observed Nature: terms like 'alien' and 'invasive species' reflect human values not nature 'uninterpreted'. The view that the earth is some kind of organism, existing in some harmonious balance, popularized in James Lovelock's *Gaia* (Lovelock, 1987) but inherent in many myths of nature held by environmental scientists, was long ago dismissed by philosopher Friedrich Nietzche as one of many 'aesthetic anthropomorphisms' whereby we interpret the world to be intelligible or purposeful by projecting our own characteristics on to it. As Robert Kirkman has noted, a belief in ecologism provides a kind of moral compass which points in a single direction: the health or well-being of the Whole: 'harmony or relatedness in general, however it is understood, is taken to be the highest good to which all else is to be subordinated' (Kirkman, 1997, p. 389). We have suggested here that this 'all else' includes a commitment to scientific objectivity. Yet this 'corruption' is ultimately likely to be futile, since Kirkman (1997, p. 389) suggests 'holistic harmony' is illusory, and 'any such attempt to orient ourselves in the universe is nothing more than a projection of ourselves and our own desires and values onto a world which is, for all we know, fundamentally indifferent to us.'

Others do not share our desires and values, and so will project different things on to the universe. Not only is it unlikely that consensus in favour of some preferred policy will emerge, but (as the CBD experience shows) even if we can get agreement, we will find ourselves without a basis for either acting or knowing whether we have acted successfully. And in the process, we will have diminished, rather than enhanced the scientific basis for action upon which we must ultimately depend. Jeff McNeely suggested that if we can't measure species loss, we might substitute a measure of how hard we are trying. 'Measures of effort may be a more useful indicator than estimated rates of species loss. For example, we can measure the number of new protected areas being established and ask if they have sufficient budgets.' As any policy analyst will quickly respond, this commits the cardinal sin of measuring policy success by measuring inputs rather than outputs, but counting inputs is sometimes *politically* attractive, because it helps signal to constituencies that the policy provides benefits to them (Behn, 1981).

Ultimately, the key to our understanding of the virtuous corruption of virtual science lies in politics – at least in part. The indeterminacy of the science in question permits the political frames attached to science to become relatively more prominent, but it is politics that shapes those frames, and we need to understand how this happens, and account for the tendency on the Left of politics in the United States in particular to view science that does not support their preferred policies in partisan terms, yet seemingly not be aware of its own politicization of science. This tendency which has provoked a similar response from the political Right, such that (*contra* Crichton) data – or at least science – appears to have taken on Democrat or Republican tinges (all to the detriment of both science and public policy). Cultural theories of risk can provide some answers here, particularly John Adams's observation that different cultural frames become more important the more virtual risks become; that is, the less determinate the science.

CULTURAL THEORIES OF RISK

John Adams has synthesized a number of perspectives on the cultural theory of risk (Adams, 1995, pp. 33–50). This is especially relevant to our analysis, because he emphasizes the greater importance of cultural factors in determining risk perceptions under conditions when science is indeterminate. In his later work, he refers to such circumstances as 'virtual risk', and we can extend the analysis to those circumstances where the science itself is more 'virtual': where there is more need to massage data, and where the science involved depends more on computer modelling than on observational science; in other words, there is more scope for the cultural appreciation of risk to intrude into both the conduct of the science and its communication in what we have termed here 'virtuous' or 'noble cause' corruption.

Adams draws upon the work of Mary Douglas and Aaron Wildavsky (1981) in their book *Risk and Culture* to point out that we can never have certain knowledge of future risks, and different people confront future risks as if the world was a different place. This is especially so with ecosystem managers, who must make decisions about the future in the face of considerable uncertainty. Holling (1979) found patterned consistencies in the beliefs of ecosystem managers which he classified in three categories of myth, each of which assumed that nature behaved in certain ways in response to human interventions. The myth he described as 'Nature Benign' saw nature as predictable, bountiful, robust, and resilient – able to absorb disturbances with little harm and suggesting a laissez-faire management approach as appropriate. (The myth can be depicted visually by a concave surface on which a marble rests: it can be dislodged, but will soon return to stability see Figure 6. 1.)

Those who adhered to a myth of 'Nature Ephemeral' saw nature as existing in a delicate balance which was likely to be disrupted by human intervention, plunging it into catastrophic collapse. It suggests that people need to tread very lightly on the earth, and act in a careful, precautionary fashion. Those who adhere to the myth of nature labelled 'Nature Perverse/Tolerant' essentially combined the first two myths: nature is robust within limits, but large enough shocks can lead to catastrophe, suggesting an interventionist management style.

Schwarz and Thompson (1990) added a fourth myth, that of 'Nature Capricious', which sees nature as inherently unpredictable and predicates a fatalistic view of the future, and suggestive of therefore (again) a fatalistic management style. But they also then mapped the myths of nature on to a matrix of beliefs about human nature that incorporated dimensions that corresponded broadly to the dimensions by which political scientists have long distinguished traditional political ideologies: freedom and equality. In the case of Schwarz and Thompson, the 'freedom' dimension had at its extremes 'Individualized' and 'Collectivized', while the vertical axis incorporated 'Inequality' and 'Equality'. Inequality was given the alternative descriptor of 'prescribed', meaning that human behaviour was seen as being constrained by restrictions imposed by some superior authority. The equality pole was also given the additional descriptor of 'prescribing', to connote the view that there were no externally imposed rules – that people negotiated rules as they went along.

In a similar vein, Thomas Sowell (2002) describes two competing visions that shape our debates about the nature of reason, justice, equality and power. The 'constrained' vision sees human nature as unchanging (at least in the shorter term) and selfish and the 'unconstrained' vision sees human nature as malleable and perfectible. Sowell suggests that many

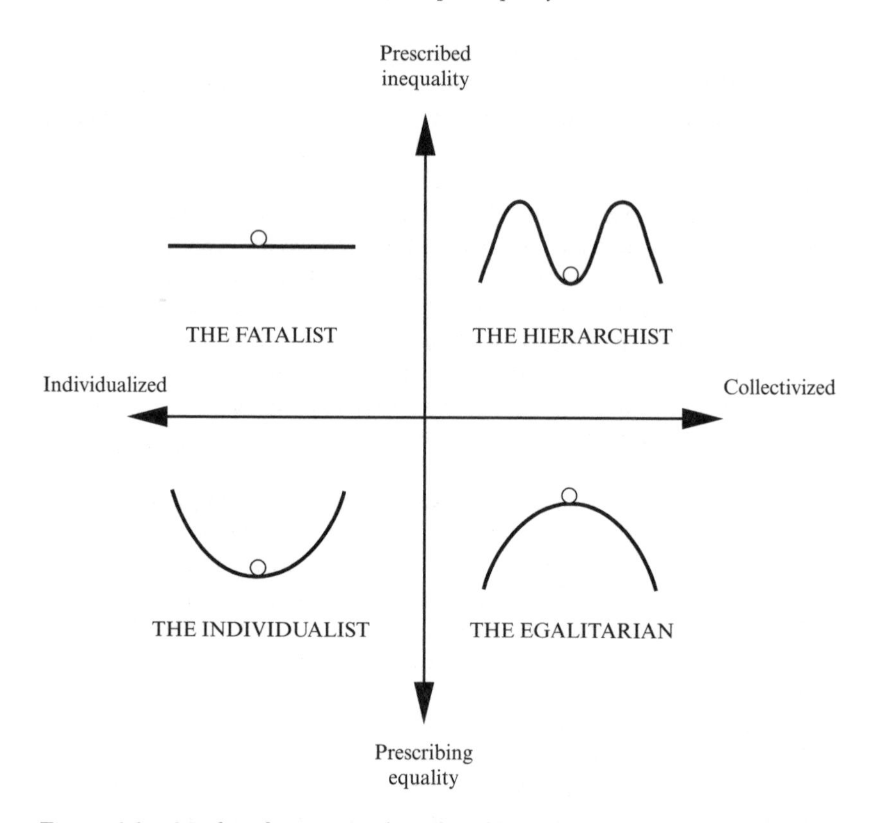

Figure 6.1 Myths of nature and myths of human nature

ethical and policy disputes often revolve around these two incompatible visions. Ehrlich and others who see humanity as having to change its culture and ethics to avert eco-catastrophe, clearly subscribe to an unconstrained vision of humanity (in Sowell's terms), and this reinforces the cultural differences that they bring to interpreting the world through indeterminate science, and sets them aside from heretics such as Lomborg (and most of the economics profession) who subscribe to a constrained vision.

This matrix mirrors closely the traditional manner in which political ideologies are distinguished. Schwarz and Thompson distinguish Individualists (high freedom, prescribing/equality), Egalitarians (prescribing/equality, collectivized), Fatalists (high freedom, prescribed/inequality), and Hierarchists (collectivized, prescribed/inequality). Liberals value individual freedom and the ability of individuals to negotiate outcomes as they go along, but they care about justice. Not surprisingly, when Schwarz and Thompson map myths of nature on to approaches to human nature, Individualists map on to 'Nature Benign'. Libertarians are represented by Fatalists (Nature Capricious) – placing a high value on individual freedom and largely indifferent to questions of distributive justice. Hierarchists map on to Nature perverse/tolerant, and with a low value placed on both freedom and equality, and in the same quadrant as conservatives and authoritarians, though any particular ideological variant would be located at different places in the quadrant. And Egalitarians are in the remaining quadrant (low freedom, high equality), which is where we would expect to find those on the traditional Left.

This suggests that social democratic ideology is more likely to resonate with the 'Nature Ephemeral' myth of nature. Social democrats are likely to carry in their heads a picture of nature existing in a fragile balance, and to continue to cling to it even when ecological science has moved on to a new paradigm involving change and perturbation, with the ultimate demise of the 'stability' paradigm being put by Donald Worster (1994) at about 1990. One might expect that the cross-pressuring of the conduct of 'normal science' would limit their ability to cling to an old myth in the face of the new paradigm, but the strong normative attachment to environmentalism provides a powerful 'force-field' to defend against inconvenient facts and theories by dismissing them as being advanced by those with less noble motives. The New Ecology cannot serve as the scientific basis for environmental policy based on the preservation of stability that is so dear to them, so they cling to outmoded scientific beliefs spun together with a normative thread.

This analysis provides the basis for understanding the types of cultural differences which people bring to matters of risk, and Adams uses them for that very purpose, to explain why people demonstrate very different reactions to what appear to be objectively speaking the same 'odds' in risk evaluation. Adams then goes on to develop an interactive theory of risk based upon these different cultures, which is beyond our purpose here. But in some of his later work he emphasizes the point that these cultural 'filters' are relatively more significant under conditions of 'virtual risk' – where the science is indeterminate – and this *is* of interest. The matrix of different cultural perspectives on risk is suggestive of how different ideological dispositions might result in different interpretations of uncertain science, and explain why data – *contra* Crichton – are more likely to take on partisan attributes when they are virtual.

There is a long-standing distinction in economics between *risk* and *uncertainty*, but (as Adams points out) all future risks are uncertain because we cannot know the future with any certainty. So the virtual nature of areas of science such as projecting future (or even present) rates of biodiversity loss or climate change makes them particularly susceptible to the projection of cultural appreciations of risk on to them. But the opportunity for projection also occurs with the construction of complex models that few others get to peer inside (and few others understand) and in the preparation of data to drive them. Virtuous corruption need not presuppose deliberate or even conscious manipulation of data or models, but simply the privileging of certain results through the lack of sufficient scepticism of data and methods that provide answers that are politically useful.

We noted in the previous chapter that cognitive science provides support for the view that the framing of science can impact upon the way in which information is processed and therefore how well individuals learn. We noted that the influence of frames in producing bias is thought to be greater when information is either incomplete or overly complex. Cultural perspectives on risk add a collective dimension to the insights of cognitive psychology, because they suggest those shared beliefs are effectively political ideologies. Mary Douglas (1992) provides a useful additional wrinkle here, in her observation that much of the contemporary politics of risk has to do with the exercise of power via the attribution of blame, and the attribution of blame is one way citizens have of countering the power of large, faceless corporations, both public and private. So fossil fuel companies are blamed for climate change, multinational chemical companies for toxic chemicals, and so on.

Frank Furedi (2001), in a provocative essay, has suggested that this 'consumer activism' (in which he includes most environmentalism) is a diversion from a more fundamental critique of capitalism, but, be that as it may, we can certainly conclude here that such blame attribution plays its part in injecting partisanship into science. As Douglas puts it:

> Blaming is a way of manning the gates through which all information has to pass. Blaming is a way of manning the gates and at the same time of arming the guard. News that is going to be accepted as true information has to be wearing a badge of loyalty to the particular regime which the person supports; the rest is suspect, deliberately censored or unconsciously ignored. (1992, p. 19)

Furedi's critique of consumer activism is essentially the same as that by the (Old) Left of what was known as the New Left, which emerged almost simultaneously with social strands such as neo-Malthusianism and the Union of Concerned Scientists at the end of the 1960s and beginning of the 1970s – the very period when Ehrlich and others were coming to prominence. The distinguishing feature of the New Left was its simultaneous emphasis on freedom and equality, so that it was often seen as being a creature of the Right (Schweitzer and Elden, 1971). But the New Left soon had a new dancing partner in the form of the New Right, which saw a similar novel combination of positions: the morality of Conservatives and the individualism of Liberals.

Grendstad and Selle (2000) argue that cultural theory is best served by treating the myths of physical and human nature as logically independent of one another, but this analysis is suggestive of a connection between the two, and at least an association between the environmentalism and the New Left which emerged together and may well continue to exert a hold over the Baby Boomer generation of scientists. Significantly, it is probably this 'New Left' emphasis on individual freedom that makes unlikely the prospect that environmentalism might travel down the dark path that is possible. This is because this suggests a deep adherence to the separation between the individual and the state which is central to Liberalism and which was notably lacking in Germany when the dark side of environmentalism was most prominent. Nevertheless, there have been warnings sounded about the implications of some anti-humanist strands in modern environmentalism since its emergence (Neuhaus, 1971).

Observer biases are well enough known in almost every area of science, and are exactly why double-blind techniques are used in designing experiments and trials. But with environmental science we have the additional problem that 'disciplines' such as conservation biology have been founded on a basis which includes certain values, and the highly political nature of the subsequent scientific endeavour allows its practitioners to exercise hegemony over an area of enquiry and defend it against external challengers. One can see a similar phenomenon with 'climate science', which is not a traditional discipline, and most of its practitioners have different disciplinary backgrounds (many are physicists) and, one suspects that those who have captured its commanding heights share a cultural disposition towards risk.

It is this tendency towards 'groupthink' that presents dangers. In other areas of science, there is a healthy contestation and science will (in the long run) be broadly self-correcting because of the various negative feedback mechanisms. With the examples of 'virtual science' we have examined here, such mechanisms (in the form of 'sceptics' or 'contrarians') are marginalized by numerous rhetorical devices such as association with 'fossil fuel interests', and so on – they do not wear the appropriate badge of loyalty, as Douglas has put it. The Lomborg affair shows the defence systems of blame and resort to conspiracy theories involving corporate interests in full flourish, with a well-developed immune system, sensitized by prior exposure to Julian Simon, resulting in the dispatch of the 'white cells' to deal with the infectious ideas of the apostate Lomborg. Electronic communications facilitated a coordinated attack, evident by the communication between and overlap among Lomborg's critics, just as text messaging permits 'swarming' of young activists to demonstrations or rioters to less politically meaningful events. The close network among the climate scientists who defended the Hockey Stick from criticism functioned similarly, and the weapons used were similar to those used against Lomborg – associating McIntyre and McKitrick with mining companies, and with ExxonMobil and questioning both their expertise and their motives as they sough to undermine 'scientific truth'.

It is not only outsiders who are subjected to attack if they stray from research that supports the preferred climate policy. But the Lomborg case revealed a deeper ideological divide through the reactions of his critics to what they regarded as his inappropriate framing of the facts: most of all, they rejected his *anthropocentrism*. This suggests that Lomborg and his critics are not just in a different ideological *quadrant*, but also in a different ideological *plane*, and this affects the way they frame facts, perceive risks, and possibly select data. By this I mean that Lomborg operates in the plane of what Theodore Lowi (1987) has called 'mainstream' approaches to regulation, where preferences are held by people in something resembling a normal distribution – perhaps skewed to the left or the right, but roughly clustered around some mean. Such regulation is usually motivated by things that are harmful in their consequences; there is little focus on whether the action in itself is right or wrong; policies can be continuously adjusted until there is a degree of benefit and cost that is acceptable to a coalition sufficient to support the measure politically.

But Lowi points out there is a different kind of regulatory politics, where the concern is not with 'harm reduction', as is the case with mainstream regulation, but with whether the action in and of itself is morally right, and this drives most of the so-called 'new politics' of which environmental politics is part, towards a radical approach to regulation. Here preferences are distributed in a strongly bipolar pattern, clustered around poles of political loyalty some considerable (and usually unbridgeable) distance apart. It is not that there is never any concern with morality in the politics of liberal regulation, nor any concern with harm reduction in radical politics, but each becomes subsumed by the other. (In fact, what Yandle (1989, 2001) has called 'Baptist-and-bootlegger' coalitions frequently form to support regulatory measures – where economic interests are helped by the strong moral arguments which disempower their interest-motivated opponents if they lack a similar moral cloak for their demands.)

So many environmentalists – and certainly Lomborg's critics – reject his anthropocentrism and instead argue for the importance of a biocentric or ecocentric approach. This helps keep the different camps apart, adds to the vehemence of the criticism, and helps environmentalists resist being coopted into mainstream politics. But the effect on the conduct of science is perhaps of more concern, because it conveys a sense of rectitude to those practising the science and marginalizes those who beg to differ with them, or even those who do not support them as enthusiastically as they might.

For example, James Hansen, despite his impeccable credentials as a climate scientist who promotes a view of warming that we should be concerned about, came in for criticism because his science-based views broke ranks with the accepted orthodoxy of radical intervention.

Hansen is widely regarded as having first sounded the climate alarm bells because his congressional testimony in 1988 really got the issue on the agenda. Indeed, his appearance at Senator Tim Wirth's Congressional hearing on climate change was itself highly politicized, and largely orchestrated by Hansen and Friends of the Earth for maximum political impact. Rafe Pomerance, then President of FOE, arranged for Hansen to testify, which he did as a private citizen to avoid the risk of censorship by his government employer. He was originally to appear before the committee in November 1987, but he convinced FOE that his testimony would not have maximum impact in the cold of autumn, and instead appeared on 23 June 1988, when Washington was sweltering in a 101°F (38°)C day. In the midst of a drought that year, Hansen's view that climate change was upon us grabbed much more attention (Boyle, n.d.).

Despite these impeccable credentials, Hansen was criticized for putting forward his 'alternative scenario'. The Hansen Alternative Scenario still calls for mitigation of carbon dioxide emissions, but pushes as higher priorities mitigation of those other greenhouse forcing agents that can be mitigated more readily, more cheaply, or with substantial co-benefits. Hansen's sin was to advocate a policy response which did not require decarbonization, and defined the problem in a way that suggested different solutions.

Hansen's is not an isolated case of a scientist being criticized for the uses to which his science or advice might be put. The whole of the IPCC exercise avoided an initial consideration of adaptation to climate change as the result of pressure from NGOs fearful that it would not send such a strong message in favour of mitigation. Roger Pielke Jr, an open-minded researcher, has reported similar pressures from colleagues. Pielke accepts that there is anthropogenic warming, but is sceptical about many claims made by those he considers alarmists, and considers it is more important to focus on developing effective policy in the face of the insoluble uncertainties in the science. In 2001, while at the National Center for Atmospheric Research at the University of Colorado, Pielke reported being pressured by colleagues to downplay the policy aspects of his work that showed that the reasons for increasing hurricane damage (both in terms of amount of physical damage and value of the damage caused) were the increased number and value of assets constructed in vulnerable areas. There was no evidence of rising frequency or intensity of hurricanes (beyond natural variability) and – while he admits that future climate change might conceivably exacerbate damage in the future – his research suggested policy action to try to mitigate climate change would be much less cost-effective than adaptation.

Pielke has an impressive list of peer-reviewed publications in leading journals (including *Science* and *Nature*) to back this view, yet he reported frequent non-scientific attempts to get him to change his views. He reported the following examples:

- A 'very prominent scientist' all but accusing him of falsifying his research results in order to hide the global warming signal in disaster losses that he '*believes must certainly be there*' [original emphasis].
- Another prominent scientist 'quite angrily and nastily' accusing him in an e-mail of being a climate change denier who refuses 'to see the truth'.
- Another prominent scientist, head of a major research unit, asking him after a lecture, not anything substantive about his lecture, but his political orientation, which the questioner had found difficulty discerning from his talk.
- The editor of a leading scientific journal asking him to 'dampen' the message of a peer-reviewed publication for fear of it being seized upon by those seeking to defend their interests in business-as-usual energy policies. Pielke wrote: 'I found this incredible – was I really being asked to change scientifically well-supported arguments based on some editorial concerns about politics?'(Pielke, 2006)

Not surprisingly (especially given the prominence of the people bringing this pressure to bear) Pielke worried that the politicization of climate science was 'reaching epidemic proportions with profound consequences for the field'. But the Lomborg case revealed a deeper ideological divide through the reactions of his critics to what they regarded as his inappropriate framing of the facts: most of all, they rejected his *anthropocentrism*.

Many environmentalists and environmental scientists see a necessity for humanity to change both its nature and its numbers to avert environmental apocalypse. The most prominent, Paul Ehrlich, perhaps best exemplifies this. Ehrlich issued a press release in August 2004 on his call for the establishment of a 'Milllennium Assessment of Human Behavior' (MAHB) which he was to issue the next day at the 89th Annual Meeting of the Ecological Society of America. Pointing to the IPCC, the conclusions of which were 'filtered somewhat by political considerations', Ehrlich noted that:

> Similarly there is a now a global effort by hundreds of scientists to evaluate the condition of the world's ecosystems – humanity's life support apparatus – called Millennium Ecosystem Assessment. But there is no parallel effort to examine and air what is known about how human cultures, and especially ethics, change, and what kinds of changes might be instigated to lessen the chances of a catastrophic global collapse. (2004)

Ehrlich saw that participation in a MAHB might cause universities to reorganize to become agents of change, especially where it was needed most: 'The need for such reorganization is most apparent in the social sciences.' He called an 'absurdity' the organization of his own university into separate departments of sociology, history, economics, political science and psychology and (heaven forbid) two departments of anthropology. (Prehistory and social anthropology are quite distinct endeavours in most universities, though sometimes co-existing in a single department.)

It is a sobering thought that Ehrlich's expectation of humanity's need to adapt its values to the dictates of nature (clearly an 'unconstrained' view of human nature in Sowell's terms) is relatively mild compared with some of the more extreme examples of ecocentrism. At the 109th meeting of the Texas Academy of Science a lecture was presented by herpetologist Professor Eric R. Pianka who railed against anthropocentrism. Pianka asked that the video camera not record his lecture, which was to mark his being named 2006 Distinguished Texas Scientist. The absence of a video recording did not prevent us from learning Pianka's views, because they had been stated previously. On a paper on his web site he stated (Pianka, n.d.):

> Humans have overpopulated the Earth and in the process have created an ideal nutritional substrate on which bacteria and viruses (microbes) will grow and prosper. We are behaving like bacteria growing on an agar plate, flourishing until natural limits are reached or until another microbe colonizes and takes over, using them as their resource.

The ample evidence presented here of a lack of objectivity in what might be broadly termed the environmental sciences, might give concern to many, because some go beyond accepting the difficulty of conducting objective science to seemingly *celebrate* bias. One author (Sarewitz, 2003) has even titled an essay 'Science and environmental policy: an excess of objectivity'. Others bemoan attempts by the media to consult dissident voices on scientific issues on the grounds that they are in a great minority and 'balance' exaggerates their credibility, ignoring the point that inexpert journalists can best allow readers test the credibility of the scientists by presenting the views of their critics. (There seems to be an expectation that the claims of 'majority' scientists should be granted a free pass – although on what basis journalists would make such a judgement is not clear.) There does, however, appear to be some awakening to the problems that this politicization might lead to.

The journal *Global Environmental Change* published an editorial by Steve Rayner in early 2006 addressing the 'widespread pathology' of using 'bad arguments for good causes' – a step in the same direction of the theme of this book, albeit one that stops short of suggesting that this might have a corrupting influence on science (Rayner, 2006). Rayner cites Lord May, formerly President of the Royal Society and Chief Scientist in the United Kingdom, erroneously exploiting the Hurricane Katrina disaster in New Orleans to try to build political support for action on climate change. As a mere Category 3 hurricane which devastated a city built below sea level, Katrina was a paradigm case of Pielke's point about vulnerability; it was a point reinforced by Cyclone Larry, a Category 5 storm that struck the coast of North Queensland in Australia in March 2006 without the loss of a single life.

Lord May had stated: 'Nobody can say that global warming played no part in the unusual ferocity of hurricanes Katrina, Rita and Wilma' (Rayner, 2006, p. 5). As Rayner remarked, 'Apart from the fact that Lord May would be the first to point out that science cannot prove a negative, both the science and economics behind the attribution of high storm damage losses to greenhouse gas emissions remain decidedly dodgy.' Roger Pielke Jr has noted the failure of scientists to correct the repeated misrepresentation of science (Pielke Jr, 2006). In the area of his expertise (assessment of disaster impacts) alone, Pielke Jr was able to point to: misrepresentation of UK flood assessments; misrepresentation by UNEP of disaster loss trends; misrepresentation by the former head of the IPCC of disaster loss trends; misrepresentation by the *New York Times* of trends in disaster losses; misrepresentation by the editor of *Science* (Donald Kennedy) of detection and attribution of trends in extreme events; misrepresentation by the editor of *Science* of attribution of Hurricane Katrina to GHG emissions; misrepresentation of literature of disaster trends and climate in an article in *Science*; misrepresentation by lead author responsible of hurricane chapter of attribution of Hurricane Katrina to GHG emissions; misrepresentation of a report on future tropical cyclone losses; misrepresentation by Al Gore of the state of hurricane science and attribution of Katrina; misrepresentation by IPCC WG II of storm surge impacts research; misrepresentation by the American Geophysical Union of seasonal hurricane forecast skill; misrepresentation by Environmental Defense of attribution of Katrina to GHGs and prospects for avoiding future hurricanes; misrepresentation in the *Washington Post* of the science of disaster trends and future impacts; misrepresentation in the Stern Report of trends in disaster losses and projections of future costs;

misrepresentation by UNEP of trends and projections in disaster losses. And this is just in Pielke Jr's area of expertise.

Lord May's statement highlights a problem in countering noble cause corruption: the lack of leadership in the scientific community. Lord May *should* know that one can't prove a negative, but the fact that he is prepared to use the 'witchcraft strategy' (of requiring witches to prove their innocence) in this example and frequently to commit the genetic fallacy, by dismissing climate sceptics as being in the pay of oil companies, indicates a belief that dishonest means can be justified by noble ends at the very highest echelons of science.

Many of the leading figures in the contemporary scientific hierarchy have credentials as environmental scientists, and many of them have connections going back a generation. In fact, while Michael Crichton might have been quite fanciful in suggesting that 'Aliens Cause Global Warming', in that this signalled the acceptance of mathematical models to make large speculations about the unknown (be it the future or the universe), his inclusion of the nuclear winter scare of the early 1980s in the historiography is entirely apposite. As well as legitimation of modelling applying data that was largely speculative or extrapolated, the nuclear winter issue saw the emergence of many of the protagonists of the politics of climate change. But he overlooked the fact that conservation biology was already leading the way towards a preference for reliance upon models rather than observational data. When we add the neo-Malthusian *Limits to Growth* study to the mix, we can see the extent to which virtual science took hold in the environmental sciences, and that the intrusion of values was there from the outset. At the time of *Limits to Growth*, the scientific establishment, such as *Nature* editor John Maddox (Maddox, 1972), were hostile to such virtual science. Now, however, its practitioners occupy the commanding heights, including the editorial desk at *Nature*.

CONCLUSION

I have argued in this book that the combination of a virtuous cause and the virtual science which has been a feature of environmental science is a dangerous one – dangerous for both the conduct of science and for the use of science as the basis of public policy-making. Public policy is almost never resolved one way or the other by some piece of scientific information, yet many scientists like to think that science will solve disputes, political disputes over questions of public policy – over climate change, over biodiversity, over genetically modified organisms, over toxic substances. But, as Mary Douglas (1992, p. 33) has noted, 'When science is used to arbitrate in these conditions, it eventually loses its independent status, and like other high priests who mix politics with ritual, finally disqualifies itself.'

The indeterminate character of much environmental science facilitates the incorporation of values into the science itself, so that we often are asked to behave in a precautionary fashion on the basis of science that already incorporates precautionary assumptions. Such mistakes carry consequences in terms of opportunity costs and wrong priorities. There are many important questions of environmental policy confronting society, but the time has come to ensure, through the insistence on better scientific practices, that we make them on the best possible science, and that we confine our arguments over matters of value to the political process. I have avoided in this book attempting to provide any personal reading of what I think the science says about biodiversity loss or climate change, and how we might approach such problems. For the record, I think both are important, but I find myself (having reviewed their conduct) less certain about the science underpinning them than I once was – as much as anything because the science cannot provide certainty. But I am more convinced both that we must have the best science at the base of policy decisions *and* that we must have it with all its limitations and uncertainties: it is the job of the journalist, not the scientist to 'simplify and exaggerate', and for the scientist to try to do this or to usurp the role of accountable political actors amounts to an improper exercise of power. Experts, as C.P. Snow put it, should be on tap, not on top.

The costs of a failure to base policy on the best possible science are immense, not least for those societies that are yet to attain levels of affluence enjoyed in the North. Two examples will suffice to make this final point: When the rail line linking Tibet to the outside world was opened in 2006, a Chinese research centre (the China Tibetology Research Centre) pressed the local population not to relinquish their practice of burning yak dung in favour of embracing the energy sources of modern society (*People's Daily Online*, 30 June 2006). Such a celebration of 'traditional' methods in the name of 'intermediate technology' ignores the point that the burning of biofuels, particularly indoors, is one of the worst environmental problems globally, and it impacts disproportionately on women and children. We need to be sure that eschewing fossil fuels (or any other energy form) is assessed on the best possible evidence, if we are not to fall foul of the problem of the second best, as economists would put it. Carbon soot, from inefficient combustion of both diesel oil and biofuels, is also an important climate forcing agent (Venkataraman *et al.*, 2005), and one which James Hansen has identified as offering better mitigation opportunities than other GHGs because doing so is easier, cheaper, and provides co-benefits by reducing health impacts.

The other example of the dangers of substituting essentially ideological solutions for science-based solutions is perhaps even more chilling: the use of tube wells in Bangladesh committed a generation of peasants to arsenic poisoning. Analysis of groundwater for use in irrigation did not include arsenic, a human carcinogen, but this analysis was later used as the basis for a decision to use these 'intermediate technology' water sources for drinking water (Pepper, 2006). Such a tragedy could have been averted if there had been a greater appreciation of the social setting, relevance and limitations of the analytical science involved.

Policy-makers can only appreciate such policy-relevant possibilities if, ironically, they put aside the political activism of scientists like Hansen. As Roger Peilke Jr put it, they need policy-relevant science that provides choices, not political science that seeks to limit them. And the choice quite properly lies with politicians, even if they make different choices to those preferred by the scientists, for whom the subjects of their expertise loom larger, removed from the larger social context and freed from the awkward realm of democratic accountability, opportunity cost and second-best problems that policy-makers must inhabit.

The efforts of scientists to usurp the role of policy-makers by alarming both them and the public is likely, ultimately, to prove futile. In doing so, they are adhering to what is known as the 'deficit model' in the 'public understanding of science', and they think (erroneously) that the public will change its mind and support the 'right' policies if only they are educated. But public understanding of science has been shown to depend crucially upon the social context in which the knowledge is at issue, with different contexts producing different views about what information is needed, by whom, and for what purpose (Irwin and Wynne, 1996). Perhaps worse still, such efforts might be counterproductive, or produce all manner of unintended consequences – a point some climate scientists now appreciate (Hulme, 2006).

Bibliography

Adam, David (2006), 'Scientists fear new attempts to undermine climate action', *The Guardian*, 21 April.

Adams, John (1995), *Risk*, London: UCL Press.

Adler, Jonathan H. (2002), 'Fables of the Cuyahoga: reconstructing a history of environmental protection', *Fordham Environmental Law Journal*, **14**, 89–146

Agerup, Martin (2003), 'Is Kyoto a good idea?', in Kendra Okonski (ed.), *Adapt or Die: The Science, Politics and Economics of Climate Change*, London: Profile Books.

Agerup, Martin (2004), 'The science behind the climate change forecasts adds up to a lot of hot air', *Daily Telegraph*, 3 May.

Alley, R.B., P.U. Clark, P. Huybrechts and I. Joughin (2005), 'Ice-sheet and sea level changes', *Science*, **310**, 456–60.

Altman, Lawrence K. and William J. Broad (2005), 'Global trend: more science, more fraud', *New York Times*, 20 December.

Anderson, Christopher (1991), 'Cholera epidemic traced to risk miscalculation', *Nature*, **354**, 255.

Anon. (1999), 'Testing, testing' (The Guardian Profile: Robert May), *The Guardian*, 30 October.

Anon. (2004), 'Keiko the whale could pose environmental threat', *ABC News Online*, 11 January, www.abc.net.au/news/newsitems/s1023640, accessed 13 January.

Anon. (2006a), 'Protection of Nessie perplexed men from the ministry', *The Herald*, 9 January, www.theherald.co.uk/news/53911, accessed 12 January.

Anon. (2006b), 'How many species inhabit the planet?', Reuters, 15 March, www.alertnet.org.thenews/newsdesk/L08736794.htm, accessed 16 March.

Antilla, Lisa (2005), 'Climate of scepticism: US newspaper coverage of the science of climate change', *Global Environmental Change*, **15**, 338–52.

Appell, David (2003), www.davidappell.com/archives/00000377.htm, accessed 4 November 2003.

Arnell, N.W., M.G.R. Cannell, M. Hulme, R.S. Kovats, J.F.B. Mitchell, R.J. Nicholls, M.L. Parry, M.T.J. Livermore and A. White (2002), 'The consequences of CO_2 stabilization for the impacts of climate change', *Climatic Change*, **53**, 413–46.

Arnold, S.F., D.M. Klotz, B.M. Collins, P.M. Vonier, P.M., L.J. Guillette Jr and J.A. McLachlan (1996), 'Synergistic activation of oestrogen receptor with combinations of environmental chemicals', *Science*, **272**, 1489–92.

Arrhenius, Svante (1896), 'On the influence of carbonic acid in the air upon the temperature of the ground', *Philosophical Magazine and Journal of Science*, **41**, 237–76.

Ashby, J., P.A. Lefevre, J. Odum, C.A. Harris, E.J. Routledge and J.P. Sumpter (1997), 'Synergy between synthetic oestrogens?', *Nature*, **385**, 494.

Auliciems, A. and I. Burton (1973), 'Trends in smoke concentrations before and after the Clean Air Act of 1956', *Atmospheric Environment*, **7**, 1063–70.

Bailey, Ronald (1995), *The True State of the Planet*, New York: Free Press.

Balint, Peter J. (2003), 'How ethics shape the policy preferences of environmental scientists', *Politics and the Life Sciences*, **22**, 14–23.

Barber, B. (1961), 'Resistance by scientists to scientific discovery', *Science*, **134**, 596–602.

Barton, Ian James and Garth William Paltridge (1984), '"Twilight at noon" overstated', *Ambio*, **13** (1), 49–51.

Bauer, Henry H. (2002), '"Pathological science" is not scientific misconduct (nor is it pathological)', *International Journal for Philosophy of Chemistry*, **8**, 5–20.

Bauerlein, Valerie (2006), 'Hurricane debate shatters civility of weather science', *Wall Street Journal*, 2 February.

Bazzaz, F., G. Ceballos, M. Davis, R. Dirzo, P.R. Ehrlich, T. Eisner, S. Levin, J.H. Lawton, J. Lubchenco, P.A. Matson, H.A. Mooney, P.H. Raven, J.E. Roughgarden, J. Sarukhan, D. Tilman, P. Vitousek, B. Walker, D.H. Wall, E.O. Wilson and G.M. Woodwell (1998), 'Ecological science and the human predicament', *Science*, **282**, 879.

Beder, Sharon (1997), *Global Spin: The Corporate Assault on Environmentalism*. Melbourne: Scribe.

Behn, Robert D. (1981), 'Policy analysis and policy politics', *Policy Analysis*, 7, 199–226.

Behringer, Wolfgang (1995), 'Weather, hunger and fear: origins of the European witch-hunts in climate, society and mentality', *German History*, **13**, 1–27.

Bellaby, Paul (2003), 'Communication and miscommunication of risk: understanding UK parents' attitudes to combined MMR vaccination', *British Medical Journal*, **327**, 725–8.

Biehl, Janet and Peter Staudenmaier (1996), *Ecofascism: Lessons from the German Experience*, Oakland, VA: AK Press.

Biel, Anders and Andreas Nilsson (2005), 'Religious values and environmental concern: harmony and detachment', *Social Science Quarterly*, **86**, 178–91.

Biosis (Biosis Research Pty Ltd) (2006), 'Wind farm collision risk for birds: cumulative risks for threatened and migratory species', report for the Australian Government Department of Environment and Heritage.

Bitman, J., H.C. Cecil, G.F. Fries (1970), 'DDT-induced inhibition of avian shell gland carbonic anhydrase', *Science*, **168,** 594–6.

Bitman, J., R.J. Lillie, H.C. Cecil, G.F. Fries (1971), 'Effect of DDT on reproductive performance of caged leghorns', *Poultry Science*, **50**, 657–9.

Boehmer-Christiansen, Sonja (1988), 'Pollution control or *Umweltschutz*?' *European Environment Review*, **2** (1), 6–10.

Bolin, B. and R.J. Charlson (1976), 'On the role of the tropospheric sulfur cycle in the shortwave radiative climate of the earth', *Ambio*, **5**, 47–54.

Botkin, Daniel B. (1990), *Discordant Harmonies: A New Ecology for the Twenty-first Century*, New York: Oxford University Press.

Botkin, Daniel B., Henrik Saxe, Miguel B. Araújo, Richard Betts, Richard H.W. Bradshaw, Tomas Cedhagen, Peter Chesson, Terry R. Dawson, Julie R. Etterson, Daniel P. Faith, Simon Ferrier, Antoine Guisan, Anja Skoldborg Hansen, David W. Hilbert, Graig Loehle, Chris Margules, Mark New, Matthew J. Sobel and David R.B. Stockwell (2007), 'Forecasting the effects of global warming on biodiversity', *BioScience*, **57**, 227–36.

Boyle, Robert H. (n.d.), 'You're getting warmer . . .', *Audubon Magazine* http://magazine.audubon.org/global.html, accessed 5 August 2006.

Brandt, J.H., M. Dioli, A. Hassanin, R.A. Melville, L.E. Olson, A. Seveau and R.M. Timm (2001), 'Debate on the authenticity of *Pseudonovibos spiralis* as a new species of wild bovid from Vietnam and Cambodia', *Journal of Zoology*, **255**, 437–44.

Branfman, Fred (2000), 'Living in shimmering disequilibrium', *Salon*, 22 April, http://dir.salon.com/story/people/feature/2000/04/22/eowilson/, accessed 4 January 2006.

Bray, D. and H. von Storch (1999), 'Climate science: an empirical example of postnormal science', *Bulletin of the American Meteorological Society*, **80**, 439–56.

Briffa, K.R. and T.J. Osborn (2002), 'Blowing hot and cold', *Science*, **295**, 2227–8.

Briffa, K.R., P.D. Jones, J.R. Pilcher and M.K. Hughes (1988), 'Reconstructing summer temperatures in northern Fennoscandinavia back to AD 1700 using tree ring data from Scots pine', *Arctic and Alpine Research*, **20**, 385–94.

Briffa, K.R., T.J. Osborn, F.H. Schweingruber, I.C. Harris, P.D. Jones, S.G. Shiyatov and E.A. Vaganov (2001), 'Low-frequency temperature variations from a northern tree-ring density network', *Journal of Geophysical Research*, **106**, 2929–41.

Briffa, K.R., T.J. Osborn, F.H. Schweingruber, P.D. Jones, S.G. Shiyatov and E.A. Vaganov (2002), 'Tree-ring width and density around the Northern Hemisphere: part 1. Local and regional climate signals', *Holocene*, **12**, 737–57.

Briffa, K.R., T.J. Osborn and F.H. Schweingruber (2004), 'Large-scale temperature inferences from tree rings: a review', *Global and Planetary Change*, **40**, 11–26.

Brumfiel, Geoff (2006), 'Academy affirms hockey-stick graph; but it criticizes the way the controversial climate result was used', *Scientific American*, 28 June, doi: 10.1038/4411032a.

Bryden, H., H. Longworth and S. Cunningham (2005), 'Slowing of the Atlantic meridional overturning circulation at 25°N', *Nature*, **438**, 655–7.

Bryson, Reid (2006), 'Why Files talks with Reid Bryson.' 19 October, www.whyfiles.org, accessed 24 October 2006.

Budiansky, Stephen (1995), *Nature's Keepers: The New Science of Nature Management*, New York: Free Press.

Budiansky, Stephen (2002), 'Diversionary tactics in environmental debate', *Nature*, **415**, 364.

Buell, Lawrence (1995), *The Environmental Imagination: Thoreau, Nature Writing, and the Formation of American Culture*, Cambridge, MA: Belknap Press.

Bürger, G. and U. Cubasch (2005), 'Are multiproxy climate reconstructions robust?', *Geophyical Research Letters*, **32**, L23711, doi:10.1029/2005GL024155.

Burgman, Mark A. (2002), 'Are listed threatened plant species actually at risk?', *Australian Journal of Botany*, **50**, 1–13.

Calabrese, Edward J. and Linda A. Baldwin (2003a), 'Hormesis: the dose–response revolution', *Annual Review of Pharmacology and Toxicology*, **43**, 175–97.

Calabrese, Edward J. and Linda A. Baldwin (2003b), 'Toxicology rethinks its central belief: hormesis demands a reappraisal of the way risks are assessed', *Nature*, **421**, 691–2.

Calamai, Peter (2004), 'Stormy weather in climate feud', *Toronto Star*, 10 February.

Cartwright, Nancy (1983), *How the Laws of Physics Lie*, Oxford: Oxford University Press.

Casellini, Nicole, Nguyen Cong Minh and Tran Lien Phong (2001), *The Integration of Economic Measures into the National Biodiversity Strategies and Action Plans of Viet Nam and South East Asia*, IUCN Regional Environmental Economics Programme, Asia and Vietnam Country Office, in association with the GEF Biodiversity Planning Support Programme, UNDP, UNEP and The World Bank.

Castles, I. (2005), post on *ClimateAudit*, 22 August.
Castles, I. and D. Henderson (2003a), 'The IPCC emission scenarios', *Energy and Environment*, **14**(2 and 3), 159–85.
Castles, I. and D. Henderson (2003b), 'Economics, emission scenarios and the work of the IPCC', *Energy and Environment*, **14**(4), 415–35.
Chapman, David S., Marshall G. Bartlett and Robert N. Harris (2004), 'Comment on "Ground vs surface air temperature trends: implications for borehole surface temperature reconstructions" by M.E. Mann and G. Schmidt', *Geophysical Research Letters*, **31**, L07205, doi: 10.1029/2003GL019054.
Chase, Alston (1987) *Playing God in Yellowstone: The Destruction of America's First National Park*, San Diego, CA: Harcourt Brace.
Chase, Alston (1995), *In a Dark Wood*, Boston, MA: Houghton Mifflin.
Cherfas, Jeremy (1994), 'How many species do we need?', *New Scientist*, (6 August), 36–40.
Chew, Matthew K., Manfred D Laubichler (2003), 'Natural enemies – metaphor or misconception?', *Science*, **301**, 52–3.
Chuine, I., P. Yiou, N. Viovy, B. Seguin, V. Daux and E. Le Roy Ladurie (2004), 'Grape ripening as a past climate indicator', *Nature*, **432**, 289–90.
Chylek, P. (2007), 'Uncertainty over weakening circulation', *Physics Today*, (March), 14.
Chylek, P., J.E. Box and G. Lesins (2004), 'Global warming and the Greenland ice sheet', *Climatic Change*, **63**, 201–21.
Claus, G. and K. Bolander (1977), *Ecological Sanity*, New York: David McKay.
Clayton, Mark (2004), 'A parking lot effect?', *Christian Science Monitor*, 5 February.
Coghlan, Andy (2003), 'Too little oil for global warming', *New Scientist*, 5 October.
Cole, Leonard A. (1983), *Politics and the Restraint of Science*, Totowa, NJ: Rowman and Allenheld.
Collins, Neil (2005), 'Global warming generates hot air', *Daily Telegraph*, 16 May, www.opinion.telegraph.co.uk/opinion/main.jhtml?xml=/opinion/2005/05/16/do1602.xml, accessed 16 May 2005.
Connor, Edward F. and Earl D. McCoy (1979), 'The statistics and biology of the species-area relationship', *American Naturalist*, **113**, 791–833.
Cook, E.R., K.R. Briffa, D.M. Meko, D.A. Graybill and G. Funkhouser (1995), 'The "segment length curse" in long tree-ring chronology development for palaeoclimatic studies', *Holocene*, **5**, 229–37.
Cook, E.R., J. Esper and R.D. d'Arrigo (2004), 'Extra-tropical Northern Hemisphere land temperature variability over the past 1000 years', *Quarternary Science Review*, **23**, 2063–74.

Cook, B.I., M.E. Mann, P. d'Odorico and T.M. Smith (2004), 'Statistical simulation of the influence of the NAO on European winter surface temperatures: applications to phenological modeling', *Journal of Geophysical Research*, **109**, D16106, doi: 10.1029/2003JD004305.

Corcoran, Terence (2002), 'An "insult to science"', *National Post*, 14 December. Available at: www.nationalpost.com/financialpost/story.html?id=%7B54B082F6-100C-4ADA-9012-EDF7EC30DD03%7D.

Covey, Curt, Stephen H. Schneider and Starley L. Thompson (1984), 'Global atmospheric effects of massive smoke injections from a nuclear war: results from general circulation model simulations', *Nature*, **308**, 21–5.

Cox, Simon and Richard Vadon (2006), 'A load of hot air?', *BBC News*, 20 April, http://news.bbc.co.uk/1/hi/magazine/4923504.stm, accessed 24 April.

Crichton, Michael (2003), 'Aliens cause global warming', Caltech Michelin Lecture, January 17.

Crichton, Michael (2005), 'Science policy in the 21st century', speech to the Joint Session AEI-Brookings Institution, Washington, DC, 25 January 2005, www.crichton-official.com/speeches/speeches_quote08.html, accessed 5 October 2005.

Crowley, T.J. (2000), 'Causes of climate change over the past 1000 years, *Science*, **289**, 270–77.

Crutzen, Paul (1995), 'My life with O_3, NO_x, and Other YZO_xs', Nobel Lecture, 8 December, nobelprize.org/chemistry/br>laureates/1995/crutzen-lecture.pdf, accessed 21 December 2005.

Crutzen, Paul J. and John W. Birks (1982), 'The atmosphere after a nuclear war: twilight at noon', *Ambio*, **11**, 114–25.

Daily, G., P. Dasgupta, B. Bolin, P. Crosson, J.D. Guerny, P. Ehrlich, C. Folke, A. Jansson, B.-O. Jansson, N. Kautsky, A. Kinzig, S. Levin, K.-G. Maler, P. Pinstrup-Anderson, D. Siniscalco and B. Walker (1998), *Food Production, Population Growth, and the Environment*, Beijer Reprint Series, Stockholm: Royal Swedish Academy of Sciences.

D'Arrigo, R.D., E.R. Cook, R.J. Wilson, R. Allan and M.E. Mann (2005), 'On the variability of ENSO over the past six centuries,' *Geophysical Research Letters*, **32**, L03711, doi: 10.1029/2004GL022055.

Dalton, Rex (2002), 'Fur flies over lynx survey's suspect samples', *Nature*, **415**, 107.

Dasgupta, Partha (2006), 'Comments on the Stern Review's Economics of Climate Change', comments prepared for a seminar on the Stern Review on the Economics of Climate Change, organised by the Foundation for Science and Technology.

Daubert (*Daubert v. Merrel Dow Pharmaceuticals, Inc.*), 113 S.Ct, 2786, (1993).

Davis, C.H.Y. Li, J.R. McConnell, M.M. Frey and E. Hanna (2005), 'Snowfall-driven growth in East Antarctic ice sheet mitigates recent sea-level rise', *Science*, **308**, 1898–901.

Davis, Robert E., Paul C. Knappenberger, Wendy M. Novicoff and Patrick J. Michaels (2003), 'Decadal changes in summer mortality in US cities', *International Journal of Biometeorology*, **47**, 166–75.

De Martino, Benedetto, Dharshan Kumaran, Ben Seymour and Raymond J. Dollan (2006), 'Frames, biases, and rational decision-making in the human brain', *Science*, **313**, 684–7.

DEFRA (2005), Changing public attitudes to climate change', DEFRA news release, 16 February.

Demeritt, David (2001), 'The construction of global warming and the politics of science', *Annals of the Association of American Geographers*, **91**, 307–37.

Deming, D. (1995), 'Climatic warming in North America: analysis of bore-hole temperatures', *Science*, **268**, 1576–7.

Deming, D. (2005), 'Global warming, the politicization of science, and Michael Crichton's "state of fear"', *Journal of Scientific Exploration* **19**(2), www.scientificexploration.org/jse.php.

Desai, Ajay and Lic Vuthy (1996), *Status and Distribution of Large Mammals in Eastern Cambodia; Results of the First Foot Surveys on Mondolkiri and Rattanakir provinces*, Flora and Fauna International, WWF, IUCN – World Conservation Union, www.undp.org.vn/projects/ras 93102/cambod/, accessed 17 July 2003.

Douglas, Mary (1966), *Purity and Danger: An Analysis of Concepts of Pollution and Taboo*, London: Routledge and Kegan Paul.

Douglas, Mary (1992) *Risk and Blame*, London: Routledge.

Edwards, J. Gordon (2004), 'DDT: a case study in scientific fraud', *Journal of American Physicians and Surgeons*, **9**(3), 83–8.

Edwards, Paul N. (1996), 'Global comprehensive models in politics and policymaking', *Climatic Change*, **32**, 149–61.

Ehrlich, A.H. and P.R. Ehrlich (1993), 'A population policy for the super-consumers', in P.H. Raven, L.R. Berg and G.B. Johnson (eds), *Environment*, Fortworth, TX: Saunders College Publishing.

Ehrlich, P.R. (1968), *The Population Bomb*, New York: Sierra Club/Ballantine Books.

Ehrlich, P.R. (1984), 'Introduction to chapter on evolution', in N. Myers (ed.), *Gaia: An Atlas of Planet Management*, New York: Doubleday.

Ehrlich, P.R. (1988), 'The loss of diversity: causes and consequences', in E.O. Wilson (ed.), *Biodiversity*, Washington DC: National Academy Press.

Ehrlich, Paul R. (with David S. Dobkin, Darryl Wheye, Stuart L. Pimm and John Kelly) (1993), *A Guide to the Natural History of the Birds of St. Lawrence Island, Alaska*, Williamsburg, VA: Center for Conservation Biology.

Ehrlich, P.R. (1995), 'Population and environmental destruction', in H.W. Kendall, K.J. Arrow, N.E. Borlaug, P.R. Ehrlich, J. Lederberg, J.I. Vargas, R.T. Watson and E.O. Wilson (eds), *Third Annual World Bank Conference on Environmentally Sustainable Development*, Washington DC: World Bank.

Ehrlich, Paul R. (1998), 'Foreword', in Peter H. Gleick (ed.), *The World's Water – The Biennial Report on Freshwater Resources*, Washington, DC: Island Press.

Ehrlich, P.R. (2000), 'The loss of population diversity and why it matters', in P.H. Raven and T. Williams (eds), *Nature and Human Society: The Quest*, Washington, DC: National Academy Press.

Ehrlich, P.R. (2000a), 'Donald Kennedy – The next editor-in-chief of *Science*', *Science*, **288**, 1349.

Ehrlich, P.R. (2001), 'The brownlash rides again', review of *The Skeptical Environmentalist: Measuring the Real State of the World* by Bjørn Lomborg, *Trends in Ecology & Evolution*, **17**, 51.

Ehrlich, Paul R. (2003), '"Clueless" ecosceptic relies on the power of wishful thinking', *Australian*, 1 October.

Ehrlich, P.R. (2004), 'Ecologist calls for creation of an international panel to assess human behavior', Stanford University press release, 1 August.

Ehrlich, Paul R. and Anne H. Ehrlich (1989), 'World's grain reserves: Lester Brown is right', *Washington Post*, 28 October.

Ehrlich, Paul and Anne Ehrlich (1996), *Betrayal of Science and Reason*, Washington, DC: Island Press.

Ehrlich, P.R. and J.P. Holdren (1975), 'Eight thousand million people by the year 2010', *Environmental Conservation*, **2**, 241–2.

Ehrlich, P.R. and J.P. Holdren (eds) (1988), *The Cassandra Conference*, College Station, TX: Texas A&M University Press.

Ehrlich, P.R. and P.H. Raven (1967), 'Butterflies and plants', *Scientific American*, **216**, 104–13.

Ehrlich, P.R. and P.H. Raven (1969), 'Differentiation of populations', *Science*, **165**, 1128–32.

Ehrlich, P.R. and S.H. Schneider (1995), 'Bets and "ecofantasies"', *Environmental Awareness*, **18** (2), 47–50.

Ehrlich, P.R., A.H. Ehrlich, and J.P. Holdren (1973), *Human Ecology: Problems and Solutions*, San Francisco, CA: W.H. Freeman and Co.

Ehrlich, P.R., A.H. Ehrlich, and J.P. Holdren (1977), *Ecoscience: Population, Resources, Environment*, San Francisco, CA: W.H. Freeman and Co.

Ehrlich, Paul R. and Edward O. Wilson (1991), 'Biodiversity studies: science and policy', *Science*, **253**, 758–62.

Ehrlich, P.R., J. Harte, M.A. Harwell, P.H. Raven, C. Sagan, G.M. Woodwell, J. Berry, E.S. Ayensu, A.H. Ehrlich, T. Eisner, S.J. Gould, H.D. Grover, R.

Herrera, R.M. May, E. Mayr, C.P. McKay, H.A. Mooney, N. Myers, D. Pimentel and J.M. Teal (1983), 'Long-term biological consequences of nuclear war', *Science*, **222**, 1283–92.

Ehrlich, P.R., C. Sagan, D. Kennedy and W.O. Roberts (1984), *The Cold and the Dark: The World After Nuclear War*, New York: W.W. Norton.

Ehrlich, Paul R., David S. Dobkin, Darryl Wheye and Stuart L. Pimm (1994), *The Birdwatcher's Handbook: A Guide to the Natural History of the Birds of Britain and Europe*, Oxford: Oxford University Press.

Ehrlich, P.R., G.C. Daily, S.C. Daily, N. Myers and J. Salzman (1997), 'No middle way on the environment', *Atlantic Monthly*, December, 98–104.

Emmanuel, K. (2005), 'Increasing destructiveness of tropical cyclones over the past 30 years', *Nature*, **401**, 665–9.

Encyclopedia Britannica (2006), 'Fatally flawed; refuting the recent study on encyclopedia accuracy by the journal *Nature*', March.

Erickson, C. (1988), 'An archaeological investigation of raised field agriculture in the Lake Titicaca Basin of Peru', doctoral dissertation, University of Illinois in Urbana.

Esper, J., E.R. Cook and F.H. Schweingruber (2002), 'Low-frequency signals in long tree-ring chronologies and the reconstruction of past temperature variability', *Science*, **295**, 2250–3, doi:10.1126/science.1066208.

European Environment Agency (1999), *The Environment in the EU at the Turn of the Century*, Luxembourg: European Environment Agency.

Falcon-Lang, Howard J. (2005), 'Global climate analysis of growth rings in woods, and its implications for deep-time paleoclimate studies', *Paleobiology*, **31**, 434–44.

Fernandez, Roberto J. (2002), 'Do humans create deserts?', *Trends in Ecology & Evolution*, **17**, 6–7.

Festinger, Leon (1962), *A Theory of Cognitive Dissonance*, Stanford, CA: Stanford University Press.

Festinger Leon, Henry W. Riecken and Stanley Schachter (1964), *When Prophecy Fails: a Social and Psychological Study of a Modern Group that Predicted the Destruction of the World*, New York: Harper & Row.

Feyerabend, Paul (1975), *Against Method: Outline of an Anarchistic Theory of Knowledge*, London: New Left Books.

Feynman, R.P. (1974), 'Cargo-cult science', *Engineering and Science*, June, 10–13.

Fauna and Flora International (n.d.), 'Conservation in the Cardamom mountains, Cambodia', www.fauna-flora.org/around_the_world/asia/forrest_protection.htm, accessed 17 July 2003.

Fierer, Noah, and Robert B. Jackson (2006) 'The diversity and biogeography of soil bacterial communities', *Proceedings of the National Academy of Sciences of the United States of America 10.1073/pnas.0507535103*, published online before print 9 January 2006.

Fog, Kåre (n.d.), 'The Lomborg story: an account of the Lomborg case', at www.lomborg-errors.dk/lomborgstory.doc, accessed 2 May 2005.

Formaini, Robert (1990), *The Myth of Scientific Public Policy*, New Brunswick, NJ: Transaction.

Frankfurt, Harry G. (2005), *On Bullshit*, Princeton, NJ: Princeton University Press.

Fu, Q., C.M. Johanson, S.G. Warren and D. Seidel (2004), 'Contribution of stratospheric cooling to satellite-inferred tropospheric temperature trends', *Nature*, **429**, 55–8.

Fumento, Michael (1993), *Science Under Siege: Balancing Technology and the Environment*, New York: William Morrow.

Funtowicz, S. and J. Ravetz (1993), 'Science for the post-normal age', *Futures*, **25**, 739–55.

Furedi, Frank (2001), 'Consuming democracy', *Spiked*, 8 June, accessed 15 August, 2005, at www.spiked-online.com/Articles/00000002DIIC.htm.

Galbreath, G.J. and R.A. Melville (2003), '*Pseudonovibos spiralis*: epitaph', *Journal of Zoology (London)*, **259**, 169–70.

Galbreath, G.J., J.C. Mordocq and F.H. Weiler (2006), 'Genetically solving a zoological mystery: was the kouprey (*Bos sauveli*) a feral hybrid?', *Journal of Zoology (London)*, **270**, 561–4.

Gale Research (1998), *Encyclopedia of Associations*, Detroit, MI: Gale Research.

Gee, Henry (2001), 'The strange case of the spiral-horned ox', *Nature*, 12 March, www.nature.com/nsu/010315/010315-4.html, accessed 28 January 2004.

Gelber, Steven M. and Martin L. Cook (1990), *Saving the Earth: The History of a Middle-Class Millenarian Movement*, Berkeley, CA: University of California Press.

Gelbspan, Ross (1998), *The Heat is On*, New York: Perseus Books.

Gelbspan, Ross (2005), *Boiling Point: How Politicians, Big Oil and Coal, Journalists and Activists Are Fueling the Climate Crisis – And What We Can Do to Avert the Disaster*, New York: Basic Books.

General Electric Co. v. Joiner. 522 US S. Ct. 136 (1997).

Giles, Jim (2005), 'Internet encyclopedias go head to head', *Nature*, 15 December, 900–901.

Gilovich, T., D.W. Griffith and D. Kahneman (eds) (2002), *Heuristics and Biases: The Psychology of Intuitive Judgment*, New York: Cambridge University Press.

Global Witness (2000), *Chainsaws Speak Louder than Words: A Briefing Document by Global Witness*, London: Global Witness.

Goklany, Indur M. (2004), 'Climate surprise: weather related mortality trends are down', 2 June, http://bmj.bmjjournals.com/cgi/eletters/328/7451/1269#1289, accessed 23 August.

Goldstein, Andrew (2002), 'Danish darts: reviled for sticking it to ecological dogma, Bjorn Lomborg laughs all the way to the bank', *Time*, 2 September, posted 18 August 2002 at www.time.com, accessed 27 January 2006.

Goodin, Robert E. (1992), *Green Political Theory*, Cambridge: Polity Press.

Grainger, Matthew (1998), 'One man's struggle to save Cambodian wildlife', *Phnom Penh Post* issue 7/14, 17–23 July, www.newspapers.com.kh/PhnomPenhPost, accessed 29 January 2004.

Greer, Germaine (1984), 'The myth of overpopulation', in Germaine Greer, *Sex and Destiny: The Politics of Human Fertility*, London: Picador.

Grendstad, G. and P. Selle (2000), 'Cultural myths of human and physical nature: integrated or separated?', *Risk Analysis*, **20**, 27–40.

Groopman, Jerome (2006), 'The preeclampsia puzzle: making sense of a mysterious pregnancy disorder', *New Yorker*, 24 July, www.newyorker.com, accessed 27 September 2006.

Grubb, Michael (2001), 'Relying on manna from heaven?', *Science*, **294**, 1285–7.

Grubler, Arnulf, Nebojsa Nakicenovic, Joe Alcamo, Ged Davis, Joergen Fenhann, Bill Have, Shunsuke Mori, Bill Pepper, Hugh Pitcher, Keywan Riahi, Hans-Holger Rogner, Emilo Lebre La Rovere, Alexi Sankovski, Michael Schlesinger, R.P. Shukla, Rob Swart, Nadejda Victor and Tae Yong Jung (2004), 'Emissions scenarios: a final response', *Energy and Environment*, **15**(1), 11–24.

Grubler, Arnulf, Brian O'Neill and Detlef van Vuuren (2006), 'Avoiding hazards of best- guess climate scenarios', *Nature*, **440**, 740.

Gwynne, Peter (1975), 'The cooling world', *Newsweek*, 28 April, p. 64.

Haas, Peter M. (2004), 'When does power listen to truth? A constructivist approach to the policy process', *Journal of European Public Policy*, **11**, 569–92.

Haber, Wolfgang (1993), 'Environmental attitudes in Germany: the transfer of scientific information into political action', in R.J. Berry (ed.), *Environmental Dilemmas: Ethics and Decisions*, London: Chapman and Hall.

Hannah, L. and Phillips, B. (2004), 'Extinction-risk coverage is worth inaccuracies', *Nature*, **430**, 141.

Hansen, James (2000), 'An open letter on global warming', 26 October, http://naturalscience.com/ns/letters/ns_let25.html, accessed 22 January 2005.

Hansen, James E. (2001), 'The forcing agents underlying climate change: an alternative scenario for climate change in the 21st century', testimony to the United States Senate Committee on Commerce, Science and Transportation, 1 May, http://www.giss.nasa.gov/research/features/senate/, accessed 22 January 2005.

Hansen, J. (2002), 'A brighter future', *Climatic Change*, **52**, 435–40.

Hansen, J., M. Sato, R. Ruedy, A. Lacis and V. Oinas (2000), 'Global warming in the twenty-first century: an alternative scenario', *Proceedings of the National Academy of Sciences*, **97**, 9875–80.

Hanson, Victor David (2006), 'The new old eco-pessimism', *American Spectator*, 15 October, http://victorhanson.com/articles/hanson101506.html, accessed 17 October.

Harrington, G.N. and K.D. Sanderson (1994), 'Recent contraction of wet sclerophyll forest in the wet tropics of Queensland due to invasion by rainforest', *Pacific Conservation Biology*, **1**, 319–27.

Harrison, Chris (2003), contribution to a symposium on 'The politicization of science' at a conference held by the American Association for Advancement of Science on 16 February, reprinted in *Politiken* on 11 March.

Harrison, Chris (2004), 'Peer review, politics and pluralism', *Environmental Science and Policy*, **7**, 357–68.

Hart, David M. and David G. Victor (1993), 'Scientific elites and the making of US policy for climate change research', *Social Studies of Science*, **23**, 643–80.

Harvey, Jeff (2001), 'The natural economy', *Nature*, **413**, 463.

Harvey, Jeff (2002), post by 'Harvey, Jeff' harvey@cto.nioo.knaw.nl at ecol-econ@csf.colorado.edu, 18 January, accessed 23 November 2005.

Hassanin, A., A. Seveau, H. Thomas, H. Bocherens, D. Billiou and B.X. Nguyen (2001), 'Evidence from DNA that the mysterious 'linh duong' (*Pseudonovibos spiralis*) is not a new bovid', *Comptes Rendus de l'Académie des Science Serie III Sciences de la Vie*, **324**, 71–80.

Hayes, Bernadette C. and Manussos Marangudakis (2001), 'Religion and attitudes towards nature in Britain', *British Journal of Sociology*, **52**, 139–55.

Heckenberger, Michael J., Afukaka Kuikuro, Urissapa Tabata Kuikuro, J. Christian Russell, Morgan Schmidt, Carlos Fausto and Bruna Franchetto (2003), 'Amazonia 1492: pristine forest or cultural parkland?', *Science*, **301**(5640), 1710.

Herf, Jeffrey (1984), *Reactionary Modernism: Technology, Culture, and Politics in Weimar and the Third Reich*, Cambridge: Cambridge University Press.

Hespe, Michelle (2002), 'Dilemma of the horns', *Weekend Australian Magazine*, 14–15 December, p. 14.

Heywood, Vernon H. and S.N. Stuart (1992), 'Species extinctions in tropical forests', in T.C. Whitmore and J.A. Sayer (eds), *Tropical Deforestation and Species Extinction*, London: Chapman & Hall.

Hirshfeld, A.N. (1996), 'Book review of *Our Stolen Future*', *Science*, **272**, 1444–5.

Holdren, John and Paul Ehrlich (eds) (1971), *Global Ecology*, New York: Harcourt Brace Jovanovich.

Holdren, J.P. and P.R. Ehrlich (1974), 'Human population and the global environment', *American Scientist*, **62**, 282–92.

Holdren, J.P., G.C. Daily and P.R. Ehrlich (1995), 'The meaning of sustainability: biogeophysical aspects', in M. Munasinghe and W. Shearer (eds), *Defining and Measuring Sustainability: The Biogeophysical Foundations*, Washington, DC: World Bank.

Holling, C.S. (1979), 'Myths of ecological stability', in G. Smart and W. Stanbury (eds), *Studies in Crisis Management*, Montreal: Butterworth.

House of Lords Select Committee on Economic Affairs (2005), 'The economics of climate change' minutes of evidence, 1 February, p. 52, www.publications.parliament.uk/pa/Id200506/Idselect/Ideconaf/12/12i.pdf, accessed 12 January, 2006.

Huang, Shaopeng, Henry N. Pollack and Po Yu Shen (1997), 'Late quaternary temperature changes seen in worldwide continental heat flow measurements', *Geophysical Research Letters*, **24**, 1947–50.

Huang, Shaopeng, Henry N. Pollack and Po Yu Shen (2000), 'Temperature trends over the past five centuries reconstructed from borehole temperatures', *Nature*, **403**, 756–8.

Hulme, Mike (2006), 'Chaotic world of climate truth', *Viewpoint*, BBC News, 4 November, http://news.bbc.co.uk/go/pr/fr/-/hi/science/nature/6115644.stm, accessed 12 December 2006.

Hwang, Woo Suk, Young June Ryu, Jong Hyuk Park, Eul Soon Park, Eu Gene Lee, Ja Min Koo, Hyun Yong Jeon, Byeong Chun Lee, Sung Keun Kang, Sun Jong Kim, Curie Ahn, Jung Hye Hwang, Ky Young Park, Jose B. Cibelli and Shin Yong Moon (2004), 'Evidence of a pluripotent human embryonic stem cell line derived from a cloned blastocyst', *Science*, **303**, 1669–74.

Hwang, Woo Suk, Sung Il Roh, Byeong Chun Lee, Sung Keun Kang, Dae Kee Kwon, Sue Kim, Sun Jong Kim, Sun Woo Park, Hee Sun Kwon, Chang Kyu Lee, Jung Bok Lee, Jin Mee Kim, Curie Ahn, Sun Ha Paek, Sang Sik Chang, Jung Jin Koo, Hyun Soo Yoon, Jung Hye Hwang, Youn Young Hwang, Ye Soo Park, Sun Kyung Oh, Hee Sun Kim, Jong Hyuk Park, Shin Yong Moon, and Gerald Schatten (2005), 'Patient-specific embryonic stem cells derived from human blastocysts', *Science*, **308**, 1777–83.

Idso, Sherwood B. (1984), 'Calibrations for nuclear winter', *Nature*, **312**, 407.

Idso, S.B. (1986), 'Nuclear winter and the greenhouse effect', *Nature*, **321**, 122.

Interlandi, Jeneen (2006), 'An unwelcome discovery', *New York Times*, 22 October.

Ioannidis, John P.A. (2005), 'Why most published research findings are false', *PLoS Med* **2**(8), e124 doi:10.1371/journal.pmed.0020124.

IPCC (2003), 'IPCC Press information on AR4 and emissions scenarios', IPCC press release, Milan, 8 December, http://www.ipcc.ch/press/pro8122003.htm, accessed 14 May 2004.

Irwin, Alan and Brian Wynne (eds) (1996), *Misunderstanding Science? The Public Reconstruction of Science and Technology*, Cambridge: Cambridge University Press.

Irwin, Paul G. (2004), 'By saving Keiko, we save ourselves' – President of the Humane Society of the United States, www.hsus.org/ace, accessed 5 July 2004.

IUCN (2003), *Red List*, www.iucnredlist.org/search/details.php?species=18576, accessed 1 December 2003.

Janis, Irving L. (1982), *Groupthink: Psychological Studies of Policy Decisions and Fiascos*, Boston, MA: Houghton Mifflin.

Jasanoff, S. (1987), 'Contested boundaries in policy-relevant science', *Social Studies of Science*, **17**, 195–230.

Jasanoff, S. (1990), *The Fifth Branch: Science Advisors as Policymakers*, Cambridge, MA: Harvard University Press.

Jasanoff, S. (1996), 'Beyond epistemology: relativism and engagement in the politics of science', *Social Studies of Science*, **26**, 393–418.

Jasanoff, S. and B. Wynne (1998), 'Science and decision making', in S. Rayner and E. Malone (eds), *Human Choice and Climate Change, vol 1. The Societal Framework*, Columbus, OH: Battelle Press.

Jastrow, Robert, William Nierenberg and Frederick W. Seitz (1990), *Scientific Perspectives on the Greenhouse Problem*, Ottawa, IL: Marshall Press, Jameson Books.

Johannessen, O.M., K. Khvorostovsky, M.W. Miles and L.P. Bobylev (2005), 'Recent ice-sheet growth in the interior of Greenland', *Sciencexpress* (20 October), www.scienceexpress.org.

Jones, Lisa (2003), '"Facts compute, but they don't convert": biologist Michael Soulé speaks from the heart – interview', *Sierra Magazine*, July-August, www.findarticles.com/p/articles/mi_m1525/iss_4_88/ai_104, accessed 3 March 2006.

Jones, P.D. and M.E. Mann (2004), 'Climate over the past millennia,' *Reviews of Geophysics*, **42**, RG2002.

Jones, P.D. and A. Moberg (2003), 'Hemispheric and large-scale surface air temperature variations: an extensive revision and an update to 2001', *Journal of Climate*, **16**(2), 206–23.

Jones, P.D., K.R. Briffa, T.P. Barnett and S.F.B. Tett (1998), 'High-resolution palaeoclimatic records for the last millennium: interpretation, integration and comparison with general circulation model control-run temperatures', *Holocene*, **8**, 455–71.

Jones, P.D., K.R. Briffa and T.J. Osborn (2003), 'Changes in the northern hemisphere annual cycle: implications for paleoclimatology?', *Journal of Geophysical Research*, **108**, (D18), 4588, doi:10.1029/2003JD003695.

Kaiser, J. (1997), 'Synergy paper questioned at toxicology meeting', *Science*, **275**, 1879.

Kalnay, Eugenia and Ming Cal (2003), 'Impact of urbanization and land-use change on climate', *Nature*, **423**, 528–31.

Kamrin, M.A. (1996), 'Editorial review of *Our Stolen Future*', *Scientific American*, **275**, 178–9.

Kapp, Bill and Kevin Ramsey (2003), 'Operation Snowplow: tigers bought, sold and killed on lucrative private market', *International Game Warden* (Spring), 18–25.

Keatinge, W.R., G.C. Donaldson, E. Cordoli, M. Martinelli, A.E. Kunst, J.P. Mackenbach, S. Nayha and I. Vuori (2000), 'Heat related mortality in warm and cold regions of Europe: observational study', *British Medical Journal*, **321**, 670–3.

Keenan, D.J. (2007), 'Grape harvest dates are poor indicators of summer warmth', *Theoretical and Applied Climatology*, **87**, 255–6.

Kelman, Steven (1981), *Regulating America, Regulating Sweden: A Comparative Study of Occupational Health and Safety*, Cambridge, MA: MIT Press.

Kellow, Aynsley (1996), *Transforming Power: The Politics of Electricity Planning*, Cambridge: Cambridge University Press.

Kellow, Aynsley (2005), 'The greenhouse and the garbage can: uncertainty and problem construction in climate policy', occasional paper 2/2005, policy paper no. 3 in Academy of Social Sciences in Australia, *Uncertainty and Climate Change: The Challenge for Policy*.

Kendall, W., K.J. Arrow, N.E. Borlaug, P.R. Ehrlich, J. Lederberg, J.I. Vargas, R.T. Watson, and E.O. Wilson (eds) (1995), *Third Annual World Bank Conference on Environmentally Sustainable Development*, Washington DC: World Bank.

Kennedy, D. (2003), 'Well, they were doing it too', *Science*, **302**,17.

Kennedy, Donald (2006), 'New year, new look, old problem', *Science*, **311**, 15.

Kennedy, Donald (2006a), 'The new gag rules', *Science*, **311**, 917.

Kennedy, D., D. Holloway, E. Weinthal, W. Falcon, P. Ehrlich, R. Naylor, M. May, S. Schneider, S. Fetter and J.-S. Choi (1998), *Environmental Quality and Regional Conflict*, New York: Carnegie Commission on Preventing Deadly Conflict.

Kennedy, D., P. Ehrlich and S. Schneider (1999), 'Professors urge change', *The Stanford Daily*, 30 March, www.stanforddaily.com/article/1999/3/30/letters, accessed 24 August, 2006.

Kiehl, Jeffrey T. (1992), 'Atmospheric general circulation modelling', in K.E. Trenberth (ed.), *Climate System Modeling*, Cambridge: Cambridge University Press.

Killingsworth, M. Jimmy and Jacqueline S. Palmer (1996), 'Millennial ecology: the apocalyptic narrative from *Silent Spring* to global warming', in Carl George Herndl and Stuart C. Brown (eds), *Green Culture: Environmental Rhetoric in Contemporary America*, Madison, WI: University of Wisconsin Press.

Kimchhay, Heng, Ouk Kimson, Kry Masphal, Sin Polin, Uch Seiha and H. Weiler (1998), *The distribution of tiger, leopard, elephant and wild cattle (gaur, banteng, buffalo, Khting Vor and kouprey) in Cambodia*, Wildlife Protection Office interim report, July, Phnom Penh, Cambodia.

Kinne, Otto (2003), 'Climate research: an article unleashed worldwide storms', *Climate Research*, **24**, 197–8.

Kirkman, Robert (1997), 'Why ecology cannot be all things to all people: the "adaptive radiation" of scientific concepts', *Environmental Ethics*, **18**, 375–90.

Kleinig, John (2002), 'Rethinking noble cause corruption', *International Journal of Police Science & Management*, **4**, 287–314.

Knudtson, Peter and David T. Suzuki (1992), *Wisdom of the Elders*, North Sydney: Allen & Unwin.

Kuhn, T.S. (1962), *The Structure of Scientific Revolutions*, Chicago: University of Chicago Press.

Kumho Tire Company, Ltd., et al., *v Patrick Carmichael*, et al., US 137 (1999) Supreme Court of the United States No. 97-1709 argued 7 December 1998 – decided 13 March 1999.

Kysar, Douglas A. and James Saltzman (2003), 'Environmental tribalism', *Minnesota Law Review*, **87**, 1092–125.

La Marca, Enrique, Karen R. Lips, Stefan Lötters, Robert Puschendorf, Roberto Ibáñez, José Vicente Rueda-Almonacid, Rainer Schulte, Christian Marty, Fernando Castro, Jesús Manzanilla-Puppo, Juan Elías García-Pérez, Federico Bolaños, Gerardo Chaves, J. Alan Pounds, Eduardo Toral and Bruce E. Young (2005), 'Catastrophic population declines and extinctions in neotropical harlequin frogs (*Bufonidae: Atelopus*)', *Biotropica*, **37**, 190–201.

Lachenbruch, A.H. and B.V. Marshall (1986), 'Changing climate: geothermal evidence from permafrost in the Alaskan Arctic', *Science*, **234**, 689–96.

Ladle, Richard J., Paul Jepson, Miguel B. Araújo and Robert J. Whittaker (2004), 'Dangers of crying wolf over risk of extinctions', *Nature*, **428**, doi:10.1038/428799b.

LaFollette, Marcel C. (1992), *Stealing into Print: Fraud, Plagiarism, and Misconduct in Scientific Publishing*, Berkeley, CA: University of California Press.

Lahsen, Myanna (2005), 'Seductive simulations? Uncertainty distribution around climate models', *Social Studies of Science*, **35**, 895–922.

The Lancet (2005), 'What is the Royal Society for?', editorial, **365**, 1746.

Lande, R. (1988), 'Demographic models of the Northern Spotted Owl (*Strix occidentalis caurina*)', *Oecologia* (Berlin), **75**, 601–7.

Landsea, Christopher W., Bruce A. Harper, Karl Hoarau and John A. Knaff (2006), 'Can we detect trends in extreme tropical cyclones?', *Science*, **313**, 452–4.

Langmuir, I. (1985), 'Pathological science: scientific studies based on non-existent phenomena', *Speculations in Science and Technology*, **8**, 77–94.

Langmuir, Irving and C. Guy Suits (1960), *The Collected Works of Irving Langmuir*, Oxford: Pergamon Press.

Lee, Kelley and Richard Dodgson (2000), 'Globalization and cholera: implications for global governance', *Global Governance*, **6**(2), 213–36.

Lee, Martha (1997), 'Environmental apocalypse: the millenarian ideology of "Earth First" ', in Thomas Robbins and Susan J. Palmer (eds), *Millennium, Messiahs, and Mayhem: Contemporary Apocalyptic Movements*, New York: Routledge.

Lewis, Owen T. (2006), 'Climate change, species-area curves and the extinction crisis', *Philosophical Transactions of the Royal Society*, **361**, 163–71.

Lindzen, Richard (2006), 'Climate of fear', *Wall Street Journal*, 12 April, www.opinionjournal.com, accessed 13 April.

Lindzen, Richard S., Ming-Dah Chou and Arthur Y. Hou (2001), 'Does the Earth have an adaptive infrared iris?', *Bulletin of the American Meteorological Society*, **82**(3), 417–32.

Lomborg, Bjorn (2001), *The Skeptical Environmentalist: Measuring the Real State of the World*, Cambridge: Cambridge University Press.

Lomborg, Bjorn. (2003), 'Science myths put to the sword', *letter to the editor of Australian*, 2 October.

Lovelock, James E. (1987), *Gaia: A New Look at Life on Earth*, Oxford and New York: Oxford University Press.

Lowi, Theodore J. (1987), 'New dimensions in policy and politics', foreword in R. Tatalovich and B. Daynes (eds), *Social Regulatory Policy*, Boulder, CO: Westview Press.

M&M Project Update (2004), http://www.uoguelph.ca/~rmckitri/research/fallupdate04/update.fall04.html, September, accessed 10 October.

MacArthur, R.H. and E.O. Wilson (1967), *The Theory of Island Biogeography*, Princeton, NJ: Princeton University Press.

Maddox, John (1972), *The Doomsday Syndrome*, London: Macmillan.

Maloney, Michael T. and Gordon L. Brady (1988), 'Capital turnover and marketable pollution permits', *Journal of Law and Economics*, **31**, 203–26.

Mann, Charles C. and Mark L. Plummer (1993), 'The high cost of biodiversity', *Science*, **260**, 1868–71.

Mann, Michael E. and Gavin A. Schmidt (2003), 'Ground vs surface air temperature trends: implications for borehole surface temperature reconstructions', *Geophysical Research Letters*, **31**, L07205, doi: 10.1029/2003GL017170.

Mann, M.E., R.S. Bradley and M.K. Hughes (1998), 'Global-scale temperature patterns and climate forcing over the past six centuries', *Nature*, **392**, 779–87.

Mann, M.E., R.S. Bradley and M.K. Hughes (1999), 'Northern hemisphere temperatures during the past millennium: inferences, uncertainties and limitations', *Geophysical Research Letters*, **26**, 759–62.

Mann, M.E., C.M. Ammann, R.S. Bradley, K.R. Briffa, T.J. Crowley, P.D. Jones, M. Oppenheimer, T.J. Osborn, J.T. Overpeck, S. Rutherford, K.E. Trenberth and T.M.L. Wigley (2003a), 'On past temperatures and anomalous late-20th century warmth', *EOS Transactions AGU*, **84**(27), 256–8.

Mann, M., R. Bradley and M. Hughes (2003b), note on paper by McIntyre and McKitrick in *Energy and Environment*, www.cru.uea.ac.uk/~timo/paleo/EandEPaperProblem_03nov03.pd.

Mann, M.E., S. Rutherford, R.S. Bradley, M.K. Hughes and F.T. Keimig (2003c), 'Optimal surface temperature reconstructions using terrestrial borehole data', *Journal of Geophysical Research*, **108**(D7), 4203, doi:10.1029/2002JD002532.

Mann, M.E., R.S. Bradley and M.K. Hughes (2004), 'Corrigendum', *Nature*, 1 July, 105.

Martin, Brian (1988), 'Nuclear winter: science and politics', *Science and Public Policy*, **15**, 321–34.

Martin, T. John (2006), letter, *Science*, **311**, 607.

Martinson, Brian C., Melissa S. Anderson and Raymond de Vries (2005), 'Scientists behaving badly', *Nature*, **435**, 737–8.

May, Robert M. (1971), 'The environmental crisis: a survey', *Search*, **2**, 122–31.

May, Robert M. (1973), *Stability and Complexity in Model Ecosystems*, Princeton, NJ: Princeton University Press.

May, Robert M. (1988), 'How many species are there on earth?', *Science*, **247**, 1441–9.

May, Robert M. (Lord May of Oxford) (2005), 'Threats to tomorrow's world', Anniversary Address, London: Royal Society.

McCook, Alison (2006), 'Is peer review broken?', *The Scientist*, **20**(2), 26.

McIntyre, Stephen (2005), post on *Climatesceptics* discussion group, 20 January.

McIntyre, Steven and Ross McKitrick (2003a), 'The M&M project: replication analysis of the Mann *et al.*, Hockey Stick', accessed 25 June 2004 at www.voguelph.ca/~rmckitri/research/trc.html.

McIntyre, Steven and Ross McKitrick (2003), 'Corrections to the Mann *et al.*, (1998), proxy data base and northern hemispheric average temperature series', *Energy and Environment*, **14**(6), 751–71.

McIntyre, Steven and Ross McKitrick (2005a), 'Hockey sticks, principal components and spurious significance', *Geophysical Research Letters*, **32**(3), L03710 10.1029/2004GL021750, 12 February.

McIntyre, Steven and Ross McKitrick (2005b), 'The M&M critique of the MBH98 northern hemisphere climate index: update and implications', *Energy and Environment*, **16**(1), 69–100.

McKitrick, Ross (2005), 'What is the 'Hockey Stick' debate about?', presentation to the APEC Study Centre, Canberra, 5 April.

McPhee, John (1971), *Encounters with the Archdruid*, New York: Farrar, Strauss & Giroux.

Meggers, Betty J. (2001), 'The continuing quest for El Dorado: round two', *Latin American Antiquity*, **12**(3), 304–25.

Meggers, Betty J. (2003), 'Revisiting Amazonia circa 1492', *Science*, **302**, 2067.

Melnick, Ronald L. (2005), 'A *Daubert* motion: a legal strategy to exclude essential scientific evidence in toxic tort legislation', *American Journal of Public Health*, **95**(S1), 30–4.

Mills, L. Scott (2002), 'False samples are not the same as blind controls', *Nature*, **415**, 417.

Mitchell, Ronald B. (1998), 'Discourse and sovereignty: interests, science, and morality in the regulation of whaling', *Global Governance*, **4**, 275–93.

Mooney, Chris (2005), *The Republican War on Science*, New York: Basic Books.

Muller, Richard (2003), 'Medieval global warming', *MIT Technology Review*, 17 December, www.technologyreview.com/articles/03/12/wo_muller 121703.asp?p=0, accessed 19 October 2004.

Muller, Richard (2004), 'Global warming bombshell', *MIT Technology Review*, www.technologyreview.com/articles/04/10/wo_muller 101504.asp

Myers, N., P.R. Ehrlich and A.H. Ehrlich (1993), 'The human population problem: as explosive as ever', in N. Polunin and J. Burnett (eds), *Surviving with the Biosphere*, Edinburgh: Edinburgh University Press.

Myers, R.A. and B. Worm (2003), 'Rapid worldwide depletion of predatory fish communities', *Nature*, **423**, 280–3.

Nakicenovic, Nebojsa, Arnulf Grübler, Stuard Gaffin, Tae Tong Jung, Tom Kram, Tsuneyuki Morita, Hugh Pitcher, Keywan Riahi, Michael Schlesinger, P.R. Shukla, Detlef van Vuuren, Ged Davis, Laurie Michaelis, Rob Swart and Nadja Victor (2003), 'IPCC SRES revisited: a response', *Energy and Environment*, **14**(2&3), 187–214.

Namias, J. (1978), 'Multiple causes of the North American abnormal winter', *Monthly Weather Review*, **106**, 279–95.

NASA (2005), 'NASA announcing 2005 is tied for warmest year on record', NASA PAO point paper, 15 December.

NASA (2006), http://www.nasa.gov/centers/goddard/news/topstory/2005/ articice, accessed 10 April.

National Research Council, Committee on Surface Temperature Reconstructions for the Last 2,000 Years (2006), *Surface Temperature Reconstructions for the Last 2,000 Years*, Washington, DC: National Research Council.

Nature (2002), 'Lynch mob turns on lynx researchers', editorial, *Nature*, **415**, 101.

Neff, Roni and Lynn R. Goldman (2005), 'Regulatory parallels to *Daubert*: influence, "sound science," and the delayed adoption of health-protective standards', *American Journal of Public Health*, **95**, S81–S91.

Neuhaus, Richard (1971), *In Defense of People: Ecology and the Seduction of Radicalism*, New York: Macmillan.

Nordhaus, William (2006), 'The *Stern Review* on the economics of climate change', unpublished paper, 17 November.

Nowell, Kristin (1999), 'Progress report to the Save the Tiger Fund', Cambodia Tiger Conservation Program, 27 August.

Oreskes, Naomi (2004), 'Science and public policy: what's proof got to do with it', *Environmental Science and Policy*, **7**, 369–83.

Oreskes, Naomi (2004a), 'The scientific consensus on climate change', *Science*, **306**, 1686.

Oreskes, Naomi (2005), 'Correction', *Science*, **307**(5708), 355.

Orofino, Suzanne (1996), '*Daubert* v. *Merrell Dow Pharmaceuticals, Inc.*: the battle over admissibility standards for scientific evidence in court', *Journal of Undergraduate Sciences*, **3**, 109–11.

Ortiz-Garcia, S., E. Ezcurra, B. Schoel, F. Acevedo, J. Soberon and A.A. Snow (2005), 'Absence of detectable transgenes in local landraces of maize in Oaxaca, Mexico (2003–2004)', *Proceedings of the National Academy of Science*, **102**, 12338–43.

Osborn, Timothy J. and Keith R. Briffa (2006), 'The spatial extent of the 20th-century warmth in the context of the past 1200 years', *Science*, **311**, 841–4.

Overpeck, J.K. Hughen, D. Hardy, R. Bradley, R. Case, M. Douglas, B. Finney, K. Gajewski, G. Jacoby, A. Jennings, S. Lamoureux, A. Lasca, G. MacDonald, J. Moore, M. Retelle, S. Smith, A. Wolfe and G. Zielinski (1997), 'Arctic environmental change of the last four centuries', *Science*, **278**, 1251–6.

Pannell, D.J. (2002), 'Review of "The Skeptical Environmentalist" by Bjorn Lomborg', *Australian Journal of Agricultural and Resource Economics*, **46**, 476–9.

Park, R.L. (2000), *Voodoo Science: The Road From Foolishness to Fraud*, New York: Oxford University Press.

Parker, David E. (2004), 'Large-scale warming is not urban', *Nature*, **432**, 290.

Patz, Jonathon A. (2004), 'Global warming: health impacts may be abrupt as well as long term', *British Medical Journal*, **328**, 1269–70.

Pearman, G.I., R.J. Charlson, T. Class, H.B. Clausen, P.J. Crutzen, T. Hughes, D.A. Peel, K.A. Rahn, J. Rudolph, U. Siegenthaler and D.S. Zardini (1989), 'Group report: what anthropogenic impacts are recorded in glaciers?', in H. Oeschger and C.C. Langway (eds), *Dahlem Workshop Reports: The Environmental Record in Glaciers and Ice Sheets*, Chichester: Wiley.

Pears Medical Encyclopaedia (1969), London: Book Club Associates.

Penny, Laura (2006), *Your Call is Important to Us: The Truth About Bullshit*, New York: Three Rivers Press.

Pepper, Daile (2006), 'Bangladeshis poisoned by arsenic sue British organisation', *The Lancet*, **367**, 9506, 199–200.

Peter, W.P. and A. Feiler (1994), 'Hörner von einer unbekannten Bovidenart aus Vietnam (Mammalia: Ruminata)', *Faunisticshe Abhandlungen staatliches Museum für Tierkunde Dresden*, **19**, 247–53.

Pianka, Eric R. (n.d.), 'What nobody wants to hear, but everybody needs to know', http://uts.cc.utexas.edu/~varanus/Everybody.html, accessed 5 April 2006.

Pielke Jr, Roger A. (2004), 'When scientists politicise science: making sense of controversy over *The Skeptical Environmentalist*', *Environmental Science and Policy*, **7**, 405–17.

Pielke Jr, Roger A. (2005), 'Misdefining "climate change": consequences for science and action', *Environmental Science & Policy*, **8**, 548–61.

Pielke, Roger A. Jr (2005a), 'Attribution of disaster losses', *Science*, **310**, 1615.

Pielke Jr, Roger A. (2005b), 'On the Hockey Stick', *Prometheus*, (July 6), http://sciencepolicy.colorado.edu/prometheus/archives/climate_change, accessed 10 January 2006.

Pielke Jr, Roger (2006), 'Looking away from misrepresentations of science in policy debate related to disasters and climate change', *Prometheus*, http://sciencepolicy.colorado.edu/prometheus/archives/climate_change, accessed 17 November.

Pielke, Roger A. Sr and Toshihisa Matsui (2005), 'Should light wind and windy nights have the same temperature trends at individual levels even if the boundary layer averaged heat content change is the same?', *Geophysical Research Letters*, **32**, L21813, doi10.1029/2005GL024407.

Pimm, Stuart and Jeff Harvey (2001), 'No need to worry about the future', *Nature*, **414**, 149–50.

Pimm, S.L., G.J. Russell, J.L. Gittleman and T.M. Brooks (1995), 'The future of biodiversity', *Science*, **269**, 347–50.

Pittock, A. Barrie (1987), *Beyond Darkness: Nuclear Winter in Australia and New Zealand*, Melbourne: Sun.

Pittock, A.B., T.P. Ackerman, P.J. Crutzen, M.C. MacCracken, C.S. Shapiro and R.P. Turco (1986), *Environmental Consequences of Nuclear War* (SCOPE 28, vol. I: *Physical and Atmospheric Effects*), Chichester: Wiley.

Plass, G.N. (1956), 'The carbon dioxide theory of climatic change', *Tellus*, **8**, 140–54.

Plater, Zygmunt (1990), 'A modern political tribalism in natural resources management', *Public Land Law Review*, **1**, 6–17.

Polacheck, Tom (2006), 'Tuna longline catch rates in the Indian Ocean: did industrial fishing result in a 90% rapid decline in the abundance of large predatory species?', *Marine Policy*, **30**, 470–82.

Pollack, H.N. and J.E. Smerdon (2004), 'Borehole climate reconstructions: spatial structure and hemispheric averages', *Journal of Geophysical Research*, **109**, D11106, doi:10.1029/2003JD004163.

Pollack, Mark A. (1997), 'Representing diffuse interests in EC policy-making', *Journal of European Public Policy*, **4**, 572–90.

Popper, Karl (1963), *Conjectures and Refutations: The Growth of Scientific Knowledge*, London: Routledge.

Popper, Karl (1968), *The Logic of Scientific Discovery*, revised edn, London: Hutchinson.

Porter, T. (1995), *Trust in Numbers*, Princeton, NJ: Princeton University Press.

Pounds, J. Alan, Michael P.L. Fogden and John H. Campbell (1999), 'Biological response to climate change on a tropical mountain', *Nature*, **398**, 611–15.

Pounds, J.A., J. Alan, Martín R. Bustamante, Luis A. Coloma, Jamie A. Consuegra, Michael P.L. Fogden, Pru N. Foster, Enrique La Marca,

Karen L. Masters, Andrés Merino-Viteri, Robert Puschendorf, Santiago R. Ron, G. Arturo Sánchez-Azofeifa, Christopher J. Still and Bruce E. Young (2006), 'Widespread amphibian extinctions from epidemic diseases driven by global warming', *Nature*, **439**, 161–7.

Proctor, Robert (1999), *The Nazi War on Cancer*, Princeton, NJ: Princeton University Press.

Raup, David M. (1991), *Extinction: Bad Genes or Bad Luck?*, New York: W.W. Norton.

Rayner, Steve (2006), 'What drives environmental policy?', *Global Environmental Change*, **16**, 4–6.

Regalado, Antonio (2006), 'Scientists' group agrees to congressional request to study temperature-history charting', *Wall Street Journal*, 10 February.

Reiter, Paul, Sarah Lathrop, Michel Bunning, Brad Biggerstaff, Daniel Singer, Tejpratap Tiwari, Laura Baber, Manuel Amador, Jaime Thirion, Jack Hayes, Calixto Seca, Jorge Mendez, Bernardo Ramirez, Jerome Robinson, Julie Rawlings, Vance Vorndam, Stephen Waterman, Duane Gubler, Gary Clark, and Edward Hayes (2003), 'Texas lifestyle limits transmission of dengue virus', *Emerging Infectious Diseases*, **9**, 86–9.

Rennie, John (2002), 'A response to Lomborg's rebuttal', *Scientific American*, ScientificAmerican.com, accessed 21 May 2004.

Reuters (2006), 'How many species inhabit the planet?', 15 March, www.alertnet.org.thenews/newsdesk/L08936794.htm, accessed 16 March 2006.

Revelle, R. and H.E. Suess (1957), 'Carbon dioxide exchange between the atmosphere and ocean and the question of an increase of atmospheric CO_2 during the past decades', *Tellus*, **9**, 18–27.

Roberts, Greg (2006), 'Creature discomforts', *The Australian*, 18 July.

Rollins, William H. (1995), 'Whose landscape? Technology, fascism, and environmentalism on the National Socialist *Autobahn*', *Annals of the Association of American Geographers*, **85**, 494–520.

Roughgarden, J. Sarukhan, D. Tilman, P. Vitousek, B. Walker, D.H. Wall, E.O. Wilson and G.M. Woodwell (1998), 'Ecological science and the human predicament', *Science*, **282**, 879.

Rutherford, S. and M.E. Mann (2004), 'Correction to "Optimal surface temperature reconstructions using terrestrial borehole data"', *Journal of Geophysical Research*, **109**, D11107, doi: 10.1029/2003JD004290.

San Francisco Chronicle (1995), 18 May.

San Francisco Chronicle (2002), 4 March.

Sarewitz, D. (2003), 'Science and environmental policy: an excess of objectivity', in R. Frodeman (ed.), *Earth Matters: The Earth Sciences, Philosophy, and the Claims of Community*, Upper Saddle River, NJ: Prentice-Hall.

Sarewitz, Daniel (2004), 'How science makes environmental controversies worse', *Environmental Science & Policy*, **7**, 385–403.

Scarrow, Howard A. (1972), 'The impact of British domestic air pollution legislation', *British Journal of Political of Science*, **2**, 261–82.

Schama, Simon (1995), *Landscape and Memory*, New York: Alfred A. Knopf.

Schell, Jonathan (1982), *The Fate of the Earth*, New York: Knopf.

Schiermeier, Quirin (2006), 'Oceans cool off in hottest years', *Nature*, **442**, 854–5.

Schmidt, Gavin A. and Michael E. Mann (2004), 'Reply to comment on "Ground vs surface air temperature trends: implications for borehole surface temperature reconstructions" by D. Chapman *et al.*', *Geophysical Research Letters*, **31**, L07206, doi: 10.1029/2003GL019144.

Schneider, Stephen H. (1989), *Global Warming: Are We Entering the Greenhouse Century?*, New York: Vintage Books.

Schneider, Stephen (1996), 'Don't bet all environmental changes will be beneficial', *APS News Online*, August, www.aps.org/aspnews/0896/11592.cfm, accessed 18 September 2006.

Schneider, Stephen H. (2001), 'A constructive deconstruction of deconstructionists: a response to Demeritt', *Annals of the Association of American Geographers*, **91**, 338–44.

Schneider, Stephen H., Starley L. Thompson and Curt Covey (1986), 'The mesosphere effects of nuclear winter', *Nature*, **320**, 491–2.

Schoenbrod, D. (2002), 'The Mau-mauing of Bjorn Lomborg', *Commentary*, September **114**(2), 51–5.

Schoenbrod, David S., and Christi Wilson (2003), 'What happened to the Skeptical Environmentalist', *New York Law School Review*, **46**, 581–614.

Schröter, Dagmar, Wolfgang Cramer, Rik Leemans, I. Colin Prentice, Miguel B. Araújo, Nigel W. Arnell, Alberte Bondeau, Harald Bugmann, Timothy R. Carter, Carlos A. Gracia, Anne C. de la Vega-Leinert, Markus Erhard, Frank Ewert, Margaret Glendining, Joanna, I. House, Susanna Kankaanpää, Richard J.T. Klein, Sandra Lavorel, Marcus Lindner, Marc J. Metzger, Jeannette Meyer, Timothy D. Mitchell, Isabelle Reginster, Mark Rounsevell, Santi Sabaté, Stephen Sitch, Ben Smith, Jo Smith, Pete Smith, Martin T. Sykes, Kirsten Thonicke, Wilfried Thuiller, Gill Tuck, Sönke Zaehle and Bärbel Zierl (2005), 'Ecosystem service supply and vulnerability to global change in Europe', *Science*, **310**, 1333–7.

Schwarz, Michiel and Michael Thompson (1990), *Divided We Stand: Re-Defining Politics, Technology and Social Choice*, Philadelphia: University of Pennsylvania Press.

Schwartz, Michael K., L. Scott Mills, Kevin. S. McKelvey, Leonard F. Ruggiero and Fred W. Allendorf (2002), 'DNA reveals high dispersal synchronizing the population dynamics of Canada lynx', *Nature*, **415**, 520–2.

Schweitzer, D.R. and J.M. Elden (1971), 'New Left as Right: convergent themes of political discontent', *Journal of Social Issues*, **27**, 141–66.

Scoones, I. (1999), 'New ecology and the social sciences: what prospects for a fruitful engagement?', *Annual Review of Anthropology*, **28**, 479–507.

Seidle, Troy (2004), 'Ideology masquerading as science: the case of endocrine disrupter screening programmes', *ATLA*, **32** (supplement 1), 669–72.

Seitz, Russell (1985), 'More on nuclear winter', *Nature*, **315**, 272.

Shackley, Simon, James Risbey, Peter Stone and Brian Wynne (1999), 'Adjusting to policy expectations in climate change modelling: an interdisciplinary study of flux adjustments in coupled atmosphere-ocean general circulation models', MIT Joint Program on the Science and Policy of Global Change report no. 48, Cambridge, MA.

Shatz, David (2004), *Peer Review: A Critical Inquiry*, Lanham, MD: Rowman & Littlefield.

Simon, Julian (1996), *The Ultimate Resource*, Princeton, NJ: Princeton University Press.

Sluijs, J.P. van der, J.C.M. van Eijnhoven, B. Wynne and S. Shackley (1998), 'Anchoring devices in science for policy: the case of consensus around climate sensitivity', *Social Studies of Science*, **28**, 291–323.

Smaglik, P. (2000), 'Climate change expert stirs new controversy', *Nature*, **407**, 7.

Smith, Daniel (2005), 'Political science', *New York Times*, 4 September.

Sokal, Alan D. (1996), 'Transgressing the boundaries: toward a transformative hermeneutics of quantum gravity', *Social Text*, **14**, 217–52.

Soon, W. and S. Baliunas (2003), 'Proxy climatic and environmental changes of the past 1000 years', *Climate Research*, **23**, 89–110.

Soulé, Michael E. (1985), 'What is conservation biology?', *BioScience*, **35**(11), 727–34.

Sowell, Thomas (2002), *A Conflict of Visions*, New York: Basic Books.

Spencer, R.W. and J.R. Christy (1992), 'Precision and radiosonde validation of satellite gridpoint temperature anomalies', *Journal of Climate*, **5**, 858–66.

Spencer, Roy (2004), 'When is global warming really a cooling', *Tech Central Station*, www.techcentralstation.com/050504H, accessed 10 May.

Der Spiegel (2004), 'Die Kurve ist Quatsch', 4 October.

Stevens, Matthew (2006), 'Pollie's parrot reveals greens' soft underbelly', *Australian*, 7 April 2006.

Stewart, Kathleen and Susan Harding (1999), 'Bad endings: American apocalypsis', *Annual Review of Anthropology*, **28**, 285–310.

Stoddard, Ed (2005), ' "Extinct" birds in comeback but no hope for dodo', *Reuters*, 12 August.

Stott, Philip (1998), 'Language for a non-equilibrium world', *Journal of Biogeography*, **25**, 1–2.

Strassel, Kimberley A. (2002), 'The missing lynx', *Wall Street Journal*, 24 January.

Suzuki, David T. and Amanda McConnell (1997), *The Sacred Balance: Rediscovering Our Place in Nature*, St Leonards, NSW: Allen & Unwin.

Tengs, T.O., M.E. Adams, J.S. Pliskin, D.G. Safran, J.E. Siegel, M.C. Weinstein, J.D. Graham (1995), 'Five-hundred life-saving interventions and their cost-effectiveness', *Risk Analysis*, **15**(3), 369–90.

Tennekes, H. (1972), 'Karl Popper and the accountability of numerical weather forecasting', *Weather*, **47**, 343–6.

Thomas, Chris D., Alison Cameron, Rhys E. Green, Michel Bakkenes, Linda J. Beaumont, Yvonne C. Collingham, Barend F.N. Erasmus, Marinez Ferreira de Siqueira, Alan Grainger, Lee Hannah, Lesley Hughes, Brian Huntley, Albert S. van Jaarsveld, Guy F. Midgley, Lera Miles, Miguel A. Ortega-Huerta, A. Townsend Peterson, Oliver L. Phillips and Stephen E. Williams (2004), 'Extinction risk from climate change', *Nature*, **427**, 145–8.

Thomas, H., A. Seveau and A. Hassanin (2001), 'The enigmatic new Indochinese bovid, *Pseudonovibos spiralis*: an extraordinary forgery', *Comptes Rendus de l'Académie des Science Serie III Sciences de la Vie*, **324**, 81–6.

Thompson, S.L., V.V. Aleksandrov, G.L. Stenchikov, S.H. Schneider, C. Covey and R.M. Chervin (1984), 'Global climatic consequences of nuclear war: simulations with three dimensional models', *Ambio*, **13**(4), 236–43.

Timm, Robert M. and John H. Brandt (2001), '*Pseudonovibos spiralis* (Artiodactyla: Bovidae): new information on this enigmatic South-east Asian ox', *Journal of Zoology*, **253**, 157–66.

Trewavas, Anthony (2001), 'Open debate is essential on conservation issues', *Nature*, **414**, 581–2.

Turco, R.P., O.B. Toon, T.P. Ackerman, J.B. Pollack and Carl Sagan (1983), 'Nuclear winter: global consequences of multiple nuclear explosions', *Science*, **222**, 1283–92.

United Nations Environment Program (2002), *Global Environmental Outlook 3*, London: Earthscan.

United States National Academy of Sciences (2006), ' "High confidence" that planet is warmest in 400 years; less confidence in temperature reconstructions prior to 1600', press release, 22 June.

United States National Research Council, Committee on Surface Temperature Reconstructions for the Last 2,000 Years (2006), *Surface*

Temperature Reconstructions for the Last 2,000 Years, Washington, DC: National Academy of Sciences.

Union of Concerned Scientists (2006), www.ucsusa.org, accessed 9 September.

University of Washington (2004), 'Say goodbye to Rudolph and other reindeer if global warming continues', press release, University of Washington, 1 December.

Usoskin, I.G., S.K. Solanki, M. Schüssler, K. Mursula and K. Alanko (2003), 'Millennium-scale sunspot number reconstruction: evidence for an unusually active sun since the 1940s', *Physical Review Letters*, **9** (211101), 1–4.

Velicogna, I. and J. Wahr (2006), 'Measurements of time-variable gravity show mass loss in Antarctica', *Scienceexpress*, 2 March.

Venkataraman, C., G. Habib, A. Eiguren-Fernandez, A.H. Miguel and S.K. Friedlander (2005), 'Residential biofuels in South Asia: carbonaceous aerosol emissions and climate impacts', *Science*, 4 March, 1454–6.

Vogel, David (1997), 'Trading up and governing across: transnational governance and environmental protection', *Journal of European Public Policy*, **4**, 556–71.

Vogel, Gretchen (2005), 'Korean team speeds up creation of cloned human stem cells', *Science*, **308**, 1096–7.

Von Storch, Hans (2006), statement to the House of Representatives Committee on Energy and Commerce hearing 'Questions surrounding the "Hockey Stick" temperature studies: implications for climate change assessments', 19 July.

Von Storch, H., E. Zorita, J.M. Jones, Y. Dimitriev, F. González-Ruoco and S.T.B. Tett (2004), 'Reconstructing past climate from noisy data', *Science*, **306**, 679–82.

Von Storch, Hans, Nico Stehr and Sheldon Ungar (2004), 'Sustainability and the issue of climate change', unpublished essay found at w3g.gkss.de/storch/Media/climate.culture.041130.pdf, accessed 13 January, 2006.

Wahl, Eugene R. and Caspar M. Ammann (forthcoming), 'Robustness of the Mann, Bradley, Hughes reconstruction of northern hemisphere surface temperatures: examination of criticisms based on the nature and processing of proxy climate evidence', *Climatic Change*, 10 May 2005, in review; 27 September 2005, revised; 12 December 2005, provisionally accepted; 28 February 2006, accepted for publication.

Wardle, David A., Lawrence R. Walker and Richard D. Bardgett (2004), 'Ecosystem properties and forest decline in contrasting long-term chronosequences', *Science*, **305**, 509–13.

Waterton, Claire and Brian Wynne (1996), 'Building the European Union: science and the cultural dimensions of environmental policy', *Journal of European Public Policy*, **3**, 421–40.

Weart, Spencer (2004), 'Reflections on the scientific process as seen in climate studies', an essay on the website *The Discovery of Global Warming*, American Institute of Physics, http://www.aip.org/history/climate, accessed 21 October 2005.

Wedgwood, C.V. (1999), *The Thirty Years War*, London: Folio Society.

Weekend Australian (2005), 30 April-1 May.

Wegman, Edward J., David W. Scott and Yasmin Said (2006), 'Ad Hoc Committee report on the "Hockey Stick" global climate reconstruction', Report for the US House Committee on Energy and Commerce.

Weiner, Jon (2005), 'Cancer, chemicals and history', *The Nation*, 7 February 2002, at: www.thenation.com, accessed 23 February.

White, Lynn Jr (1967), 'The historical roots of our ecologic crisis', *Science*, **155**, 1203–7.

Whitfield, John (2002), 'Locking horns', *Nature*, 28 February, www.nature.com/nsu/020225/020225–9.html, accessed 17 July 2003.

Wigley, Tom M.L. (convening lead author) (2006), 'Executive summary', in Thomas R. Karl, Susan J. Hassol, Christopher D. Miller, and William L. Murray (eds), *Temperature Trends in the Lower Atmosphere – Understanding and Reconciling Differences*, report by the US Climate Change Science Program and the Subcommittee on Global Change Research, April 2006.

Wigley, T.M.L. and S.C.B. Raper (2002), 'Reasons for larger warming projections in the IPCC Third Assessment Report', *Journal of Climate*, **15**, 2945–52.

Wildavsky, Aaron and Mary Douglas (1981), *Risk and Culture*, Berkley CA: University of California Press.

Wildlands Project, www.wildlandsproject.org.

Wilkinson, Todd (2002), 'Catfight ensues over case of lynx fur planted in forests', *Christian Science Monitor*, 10 January 2002, www.csmonitor.com/2002/0110/p2s2-uspo, accessed 11 January.

Wilson, E.O. (1991), *The Diversity of Life*, Cambridge, MA: Harvard University Press.

Wilson, E.O. (2006), http://www.answers.com/topic/edward-osborne-wilson, accessed 4 January.

Winner, Langdon (1977), *Autonomous Technology; Technics-out-of-control as a Theme in Political Thought*, Cambridge, MA: MIT Press.

WIT (2000), 'Cambodia's rediscovered wildlife', *WIT's World Ecology Report*, **XII** (3) (Fall) 11.

WMO (2006), Commission on Atmospheric Sciences S Tropical Meteorology Research Program, Steering Committee for Project TC-2: Scientific Assessment of Climate Change Effects on Tropical Cyclones, statement on tropical cyclones and climate change, February.

World Climate Report (2006), www.worldclimatereport.com, 11 January, accessed 13 January 2006.
Worster, Donald (1993), *The Wealth of Nature: Environmental History and the Ecological Imagination*, New York: Oxford University Press.
Worster, Donald (1994), *Nature's Economy: A History of Ecological Ideas*, Cambridge: Cambridge University Press.
WWF (n.d.), 'Southeastern Indochina dry evergreen forests (IM0210)', Wild World Profile, www.worldwildlfie.org/wildworld/profiles/terrestrial/im/im0210_full, accessed 17 July 2003.
Wynne, B. (1989), 'Sheepfarming after Chernobyl: a case study in communicating scientific information', *Environment*, **31**, 11–15, 33–9.
Yandle, Bruce (1989), *The Political Limits of Environmental Regulation*, New York: Quorum Books.
Yandle, Bruce (2001), 'Bootleggers, Baptists and global warming', in Terry L. Anderson and Henry I. Miller (eds), *The Greening of US Foreign Policy*, Stanford, CA: Hoover Press.

WEBSITES

Climate Audit www.climateaudit.org
http://davidappell.com/archives/00000497.htm, accessed 23 February 2005.
http://en.wikipedia.org/wiki/Wager_between_Julian_Simon_and_Paul_Ehlich', accessed 11 November 2005.
www.sciencepolicy.colorado.edu/prometheus, accessed 8 February 2005.
http://sciencepolicy.colorado.edu/prometheus/archives/disasters, accessed 15 March 2006.
http://www.answers.com/topic/edward-osborne-wilson, accessed 4 January 2006.
http://www.davidappell.co/archives/00000377.htm, accessed 4 November 2003.
RealClimate, www.realclimate.com

Index